Preface

This monograph has been written for sixth form science teachers with the object of giving them a broad general outline of the ways in which chemistry is affected by the atomic nucleus — so often thought of by chemists as being a heavy inert mass in the centre of the atom of no importance to them whatsoever.

The topic is a very large one and it has been quite impossible to do more than sketch it within the confines of such a short publication. What has been attempted is to give a general coverage of the subject including, in the first four chapters, the essential nuclear physics. To supplement this, there are reading lists at the end of each chapter. As far as possible the articles referred to are readily accessible reviews or are in text books and it is sincerely hoped that some readers will be inspired to read at least some of them. *Figures 1, 5,– 20, 26* and *27* are reprinted, by permission, from *Introduction to the atomic nucleus*, J. G. Cuninghame (Amsterdam: Elsevier, 1964).

It is a great pleasure to be able to acknowledge the ungrudging help which has been given me while writing this monograph by my colleagues, Drs C A Baker, D. Brown, A. T. G. Ferguson, N. R. Large and the late Dr E. R. Rae, all of the Atomic Energy Research Establishment at Harwell. They read and criticized all or part of the manuscript, and their comments and suggestions have greatly improved the final product. Finally, I am most grateful to my wife for typing both the original and final versions, and for assisting in the preparation of the manuscript in every possible way.

J. G. Cuninghame
Harwell
March 1972

CONTENTS

Part 1. Basic Nuclear Physics

1. Properties of the Atomic Nucleus

Introduction

The purpose of this chapter is to define, and describe briefly, the essential fundamental properties of the atomic nucleus so as to provide a basis for the three following chapters in which its make-up and behaviour will be examined.

Mass, charge and radius

In 1911 Rutherford produced a classic paper in which he explained the hitherto anomalous results of α-particle scattering experiments by proposing that the positive charge of an atom (which must be present to balance its negative electrons), together with most of its mass, occupied a small volume, which he called the nucleus, at the centre of the atom.

The nucleus is made up of protons and neutrons, each weighing about one atomic mass unit, one amu being one-twelfth of the mass of the isotope ^{12}C.* (An electron weighs about 1/1840 amu.) The protons are positively charged and their number constitutes the 'atomic number', Z, of the atom while the total of neutrons and protons makes up the 'mass number', A. 'Isotopes' of the same element have the same number of protons, but different numbers of neutrons. Nuclei having the same neutron number, N, are called 'isotones': those having the same value of A but a different value of Z are 'isobars', while nuclei with the same Z and A but different excitation energies, are 'isomers'.

Nuclei in their ground state are almost spherical and have an approximately constant density. Their size can therefore be characterized by their radius which is expressed in terms of A

$$r = r_0 A^{1/3} \qquad\qquad 1$$

The constant r_0 has a value of about 10^{-13} cm and so it is convenient to use 10^{-13} cm as a unit of length. It is called a fermi. The radius can be measured by studying interactions of bombarding particles with either the short range nuclear force which binds the neutrons and protons together, or with the coulomb force of the protons. Radii measured in these two ways are not quite the same because of a

---

* 1 u (amu) $\approx 1.6605 \times 10^{-27}$ kg.

difference in nature between the two forces, but which of these results is called the nuclear radius is simply a matter of definition. Radii are all of the order of a few fermi.

The nuclear angular momentum, magnetic and electric moments

In 1924 Pauli proposed that the hyperfine splitting of the fine structure lines observed in optical spectrograms was caused by the fact that the nucleus had an angular momentum which coupled with that of the orbital electrons. This angular momentum, often loosely called the 'spin' of the nucleus, is characterized by the quantum number I (or J in some notations). The associated magnetic moment of the nucleus has the value

$$\mu_I = g_I . I \qquad\qquad 2$$

where the proportionality constant g_I is called the nuclear gyromagnetic ratio. μ_I is measured in nuclear magnetons (nm). One nm is the magnetic moment which might be expected for the proton by analogy with the Bohr magneton used for electron magnetic moments, and so is given by

$$\mu_p = \frac{eh}{2m_p c} \qquad\qquad 3$$

where m_p is the proton mass and e its charge. Nuclei having atomic number, Z, and neutron number, N, both even, have $I = 0$ and consequently have no magnetic moment.

Actual measurement of the proton magnetic moment shows that it does not have the value of 1 nm which is expected from equation 3 but that

$$\mu_p = 2.79268 \pm 0.00006 \text{ nm}$$

This is because the proton is not a simple particle at all but is a complex structure. The neutron is also not simple because if it were, being uncharged, it would have zero magnetic moment. In fact its moment is

$$\mu_n = - 1.91354 \pm 0.00006 \text{ nm}$$

Finally, the deuteron magnetic moment should be found by summing those of the proton and neutron

$$2.79268 - 1.91354 = 0.87914 \text{ nm}$$

The measured value is 0.8576, the discrepancy being due to the nature of the nuclear forces present.

In addition to the magnetic moment, nuclei may also have an electric quadrupole moment and this may give rise to deviations in the hyperfine structure. Where such an electric moment exists it arises from the fact that the proton charge distribution of the nucleus is not spherical. It is given by

$$Q = \frac{4}{5}\eta Z r^2 \qquad\qquad 4$$

where η is a factor indicating the deformation of the charge

$$\eta = \frac{b - a}{r} \qquad\qquad 5$$

where a and b are the semi-axes of the ellipsoid and r the average radius $\{= (b + a)/2\}$. The largest values of η are found in the rare earths, that for ^{176}Lu $(= +0.174)$ being the highest.

Nucleon quantum numbers and the total nuclear angular momentum

The nuclear angular momentum quantum number, I (or J), is an amalgam of the quantum numbers of all the nucleons making up the nucleus and hence it is necessary to understand how these individual nucleons are quantized. A nucleon has a quantum number, j, which represents its total angular momentum. This is the vector sum of the intrinsic spin and its orbital angular momentum (a measure of the ellipticity of the orbit), represented by s and l respectively. s has a value of $\frac{1}{2}$ for neutrons, protons and electrons, and hence $j = l \pm \frac{1}{2}$). This 'strong coupling' (or 'jj-coupling') of the spin and orbital angular momenta is the fundamental postulate of the Shell model (p 11). The orientation of the orbit in space with respect to some arbitrary direction is characterized by the quantum number m_j which has $(2j + 1)$ integer spaced values from $+j$ to $-j$. Finally, there is a principal quantum number, n, which is an indicator of the total energy of the particular group of orbits. It has less meaning than the analogous quantum number used in atomic spectroscopy because the nucleons are not moving in orbits in a single coulomb field as the electrons are, but are bound together by the very strong short-range nuclear forces. There is an energy gap between the groups of orbits having different values of n, but it is very much less well defined than the corresponding gaps in electronic orbits: the energy of the orbits varies by only about a factor of 10 for a heavy nucleus, compared with a factor of about 10^4 for the electronic orbits. In nuclear quantization n is often replaced by the radial quantum number, v, which has a value of $(n - l)$.

The Pauli exclusion principle applies to nucleons and therefore it is not possible for two of them to have the same four quantum numbers, v, l, s, and m_j, although one proton and one neutron may share them. The 'state' of a nucleon is usually specified by quoting its l and j quantum numbers, together with a prefix showing the

Table 1. The hierarchy of states available to nucleons

	n = 1	n = 2		n = 3			n = 4				n = 5				
n; (+1, 2, 3, ...)	1	2	2	3	3	3	4	4	4	4	5	5	5	5	5
v = (n − l)	1	1	1	1	1	3	1	1	3	3	1	1	3	3	5
l = (n − a positive integer)	0	1	1	2	2	0	3	3	1	1	4	4	2	2	0
Level name	1s	1p	1p	1d	1d	2s	1f	1f	2p	2p	1g	1g	2d	2d	3s
j = (l ± 1/2)	1/2	3/2	1/2	5/2	3/2	1/2	7/2	5/2	3/2	1/2	9/2	7/2	5/2	3/2	1/2
State name	1s 1/2	1p 3/2	1p 1/2	1d 5/2	1d 3/2	2s 1/2	1f 7/2	1f 5/2	2p 3/2	2p 1/2	1g 9/2	1g 7/2	2d 5/2	2d 3/2	3s 1/2
m_j from (+j...−j)	1/2...−1/2	3/2...−3/2	1/2...−1/2	5/2...−5/2	3/2...−3/2	1/2...−1/2	7/2...−7/2	5/2...−5/2	3/2...−3/2	1/2...−1/2	9/2...−9/2	7/2...−7/2	5/2...−5/2	3/2...−3/2	1/2...−1/2
Total no. of neutrons or protons allowed in state	2	4	2	6	4	2	8	6	4	2	10	8	6	4	2
No. of nucleons in shell	2	6		12			8	22			32				
Magic no. (cumulative total of nucleons in shell)	2	8		20			28	50			82				

	1h 11/2	1h 9/2	2f 7/2	2f 5/2	3p 3/2	3p 1/2	1i 13/2	1i 11/2	2g 9/2	2g 7/2	3d 5/2	3d 3/2	4s 1/2	1j 15/2
n; (+1, 2, 3, ...)	6	6	6	6	6	6	7	7	7	7	7	7	7	8
$\nu = (n - l)$	1	1	3	3	5	5	1	1	3	3	5	5	7	1
$l = (n - a\ \text{positive integer})$	5	5	3	3	1	1	6	6	4	4	2	2	0	7
Level name	1h	1h	2f	2f	3p	3p	1i	1i	2g	2g	3d	3d	4s	1j
$j = (l \pm \tfrac{1}{2})$	$\tfrac{11}{2}$	$\tfrac{9}{2}$	$\tfrac{7}{2}$	$\tfrac{5}{2}$	$\tfrac{3}{2}$	$\tfrac{1}{2}$	$\tfrac{13}{2}$	$\tfrac{11}{2}$	$\tfrac{9}{2}$	$\tfrac{7}{2}$	$\tfrac{5}{2}$	$\tfrac{3}{2}$	$\tfrac{1}{2}$	$\tfrac{15}{2}$
State name	1h 11/2	1h 9/2	2f 7/2	2f 5/2	3p 3/2	3p 1/2	1i 13/2	1i 11/2	2g 9/2	2g 7/2	3d 5/2	3d 3/2	4s 1/2	1j 15/2
m_j from $(+j...-j)$	$\tfrac{11}{2}...-\tfrac{11}{2}$	$\tfrac{9}{2}...-\tfrac{9}{2}$	$\tfrac{7}{2}...-\tfrac{7}{2}$	$\tfrac{5}{2}...-\tfrac{5}{2}$	$\tfrac{3}{2}...-\tfrac{3}{2}$	$\tfrac{1}{2}...-\tfrac{1}{2}$	$\tfrac{13}{2}...-\tfrac{13}{2}$	$\tfrac{11}{2}...-\tfrac{11}{2}$	$\tfrac{9}{2}...-\tfrac{9}{2}$	$\tfrac{7}{2}...-\tfrac{7}{2}$	$\tfrac{5}{2}...-\tfrac{5}{2}$	$\tfrac{3}{2}...-\tfrac{3}{2}$	$\tfrac{1}{2}...-\tfrac{1}{2}$	$\tfrac{15}{2}...-\tfrac{15}{2}$
Total no. of neutrons or protons allowed in state	12	10	8	6	4	2	14	12	10	8	6	4	2	16
No. of nucleons in shell							44							58
Magic no. (cumulative total of nucleons in shell)							126							184

N.B. There may be proton shell closures at 114 and 164

Both protons and neutrons are arranged according to this table, but small differences in energy between corresponding proton and neutron states may result in different 'magic numbers' (p 10) emerging when the numbers of nucleons are large. Note that the same number of states exists if j–j coupling is replaced by some other scheme, but the energies of the states will be different.

order, the l quantum number being denoted by the symbols s, p, d, f — taken from atomic spectroscopy. Thus, s stands for $l = 0$, p for $l = 1$, etc. Table 1 shows how the hierarchy of nucleon states is formed.

The nucleons are arranged in the nucleus in accordance with the rules given above. The whole system then has a number of possible discrete energy states which are determined by its wave function. These are the nuclear levels and each has its own characteristic angular momentum, I, energy and parity. The level having the lowest value of I is called the ground state of the nucleus and is arbitrarily said to have zero energy. The way in which the total nuclear angular momentum, I, is obtained depends on the coupling scheme used, but however it is formed, it is the resultant of the individual momenta of all the nucleons in it. Since the nucleons all have $s = \frac{1}{2}$ and l zero or integral, nuclei in their ground state having even-A must have an integral or zero value of I whilst those with odd-A must have half-integral values. Even-A nuclei having even-Z and even-N have $I = 0$, odd-Z odd-N nuclei have integral values of I. The actual value of the angular momentum having I as its quantum number is $\hbar\sqrt{I(I + 1)}$.

Statistics and parity

In the wave mechanical representation of a system such as a nucleus, the particles in the system have three space coordinates and one spin coordinate. Such systems can be divided into two broad classes according to whether their wave function changes sign (i.e. is antisymmetric) or not when all the coordinates of two equal particles are exchanged. If the wave function does change sign the system follows a type of non-classical statistics called Fermi–Dirac and the particles are known as fermions. They include the proton, neutron and electron and obey the Pauli exclusion principle. Systems of the other class have Bose–Einstein statistics and the particles are called bosons. They include the α-particle and photon and do not obey the Pauli principle.

If a nucleus has a wave function, ψ, which has space coordinates (x, y, z) and these coordinates are reflected through the origin, the wave function may or may not change sign.

$$\psi(x, y, z) = \psi(-x, -y, -z)$$
or
$$\psi(x, y, z) = -\psi(-x, -y, -z)$$

In the first case the nucleus is said to have even parity, in the second it has odd. Parity and orbital angular momentum are related, so that parity is even for even l and odd for odd l. It is normally conserved during the interactions of nuclei except in a few cases such as that of β-decay. It has an important effect on whether transitions between nuclear levels are allowed or not.

Suggestions for further reading

R. E. Peirls, 'Nuclear matter', *Endeavour*, 1963, **22**, 146.

B. L. Cohen, 'Nuclear orbital structure', *Int. Sci. Technol.*, 1963, Nov., 65.

D. R. Bés and Z. Szymanski, 'The shape of the atomic nucleus', *Sci. Prog., Oxf.*, 1967, **55**, 187.

M. Baranger and R. A. Sorensen, 'The size and shape of atomic nuclei' *Scient. Am.*, 1969, Aug., 59.

G. T. Garvey, 'Nuclear mass relations', *A. Rev. nucl. Sci.*, 1969, **19**, 433.

2. Construction of the Nucleus

The nuclear force

The idea of a nucleus composed of neutrons and protons confined within a small volume requires that some force with properties quite different from those of the electrostatic (coulomb) force must be operating. In the first place the protons are positively charged and so the force must attract them to each other so as to nullify their mutual electrostatic repulsion, and secondly it must also act on the neutrons. However, it must not be a simple attractive force or the whole nucleus would collapse, and it must therefore become repulsive at very short range. Furthermore, it was shown in Chapter 1 (p 1) that nuclei do have a very definite radius of about 10^{-13} cm and this immediately suggests that the nuclear force has only a short range which is of about this order. Within this range, however, the force is extremely strong, being about 10^2 times as strong as the electrostatic force and 10^{38} times as strong as gravity. Another crucial piece of evidence is illustrated in *Fig. 1* in which the binding fraction, B/A, of nuclei is plotted as a function of the mass number, A. B is the total energy binding the whole nucleus together, *i.e.* the difference between the mass of the nucleus and the sum of the masses of the

FIG. 1. The binding fraction B/A in MeV per atomic mass unit as a function of A. The inset is an enlargement of the first part of the curve and shows the increased stability of nuclei having a closed $4n$ sub-shell.

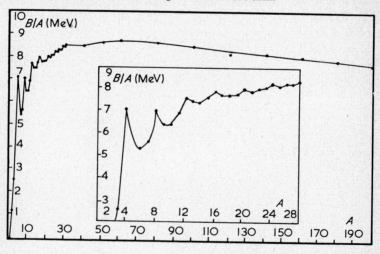

individual nucleons expressed in energy units (p 17), and the binding fraction is therefore the average binding energy of the individual nucleons. As can be seen, B/A is nearly constant with A except in the special case of the very light nuclei, and this fact suggests that the force is saturated in much the same way as the molecular binding force in a liquid is saturated.

Basically, there are three possible types of force between a pair of nucleons; neutron–proton (np), proton–proton (pp) and neutron–neutron (nn). In addition, the spins of the two nucleons can be either parallel (triplet case, total spin = 1) or anti-parallel (singlet case, total spin = 0). Thus there are triplet and singlet forces, written for example ^{3}(np) and ^{1}(np), and similarly for the (nn) and (pp) forces. Experimental evidence shows that, provided they are either both singlet or both triplet forces, the (pp) and (nn) forces are equal: this is called 'charge symmetry of nuclear forces'. The wider principle, that the (np) forces are also identical ('charge independence of nuclear forces') is true for some of the possible quantum states of a neutron–proton pair, but not for all. The evidence also shows that singlet forces are not equal to triplet forces. This is 'spin dependence of nuclear forces'.

The origin of the nuclear force lies in the idea, first suggested by Yukawa in 1935, of π-meson exchange. This exchange force can be pictured in a rather crude way by imagining that the (np) force results from the emission by protons of π^+-mesons which are absorbed by neutrons, or by neutrons emitting π^--mesons which are absorbed by protons. In a similar way (pp) and (nn) forces result from the exchange of π^0-mesons between the pairs of like nucleons. While the current theoretical view of the nuclear force is certainly far more sophisticated than this simple picture, it is still not capable of explaining all the existing experimental results.

Potential wells and barriers

Any particle approaching a nucleus is subjected to a force which is manifested as a nuclear potential whose shape depends on the nature of the various forces involved. *Figure 2* illustrates the shape of the potential for the cases of a neutron or a proton interacting with a heavy nucleus. The neutron is uncharged and at a distance from the centre of the nucleus $r = R$, where R is the sum of the radii of the nucleus and the neutron, the attractive nuclear force suddenly comes into play and the neutron is absorbed by the nucleus ('falls into the potential well'). In the case of the proton the picture is complicated by the fact that it first experiences the repulsive coulomb force and so the potential rises until R is reached, ('the proton climbs over the potential barrier'), when the much stronger nuclear force takes over and once more the nucleon falls into the potential well.

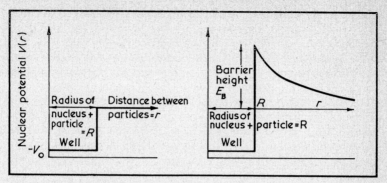

FIG. 2. The square well nuclear potential. The figures show how the potential varies with distance for a neutron (left hand diagram) or a proton interacting with a nucleus. The figures are purely diagrammatic and do not attempt to represent the square well potential to scale.

The exact shape of the potential depends on the assumptions made about the forces. The shape illustrated in *Fig. 2* is called the square well potential and is merely a convenient theoretical fiction. A more likely potential shape for the proton interaction is shown in *Fig. 3*.

Models and magic numbers

In an ideal world it would be possible to calculate how a nucleus should behave in any particular circumstances. In practice a mathematical interpretation of a system containing a number of bodies (the nucleons) interacting by means of a force which is only partially understood (the nuclear force), simply cannot be achieved. To overcome this problem, models are used in which systems are chosen

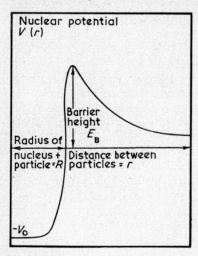

FIG. 3. A more likely shape for the potential in the case of a proton interacting with a nucleus.

which resemble the nucleus as closely as possible in the particular aspects under investigation, but which are simple enough for mathematical analysis of their properties to be made. The results of this analysis are then applied to the real nucleus.

There are many different nuclear models, but they can be broadly divided into two main classes which make exactly opposite basic assumptions. In one class, exemplified by the Liquid Drop model, there is assumed to be a very strong interaction between all the nucleons, and the nuclear properties depend on this interaction. On the other hand, the Independent Particle models, whose best known representative is the Shell model, assume that each nucleon moves independently of the others in a nuclear potential which derives from all the nucleons. The neutrons and protons are arranged in pairs having opposite spin and the nuclear properties depend only on one or two unpaired nucleons in the highest energy levels.

The Liquid Drop model, first described by N. Bohr and F. Kalckar in 1937, treats the nucleus as a drop of an incompressible liquid with a uniform electric charge. The resultant nuclear force can then be regarded as the surface tension and it decreases very sharply once the nuclear radius is exceeded. It is opposed by the coulomb force of the protons within the nucleus which falls off much more slowly with increase in radius. Under normal conditions the nuclear force far outweighs the coulomb force and the nucleus is stable, but in some circumstances, such as fission (p 35), the nucleus stretches out until the two forces just become equal. At this point a slight increase in radius will allow the coulomb force to overcome the surface tension and fission takes place. It is in the description of such collective nuclear processes as fission that the model is most successful. Its main disadvantage is that the calculations are not particularly simple, and become even harder when refinements such as compressibility, application to highly distorted drop shapes *etc.* are made, so as to make it resemble the real nucleus more closely.

Independent Particle models were developed in the early 1930s, but when the Liquid Drop model appeared in 1937 its immediate attractiveness eclipsed them. However, in 1948 Maria Mayer suggested that there was a large body of experimental evidence which seemed to indicate the existence of closed shells of nucleons analogous to the extra-nuclear electron shells. In the Shell model it is assumed that the spin (s) and orbital angular momentum (l) of any nucleon are coupled ('-jj-coupling', p 3) and this results in the splitting of the level into two states. If the two momenta are parallel the resulting state has $j = l + s$ and lies lower in energy than the $j = l - s$ state in which they are anti-parallel. The nuclear potential of the system is given a shape which depends on the assumptions

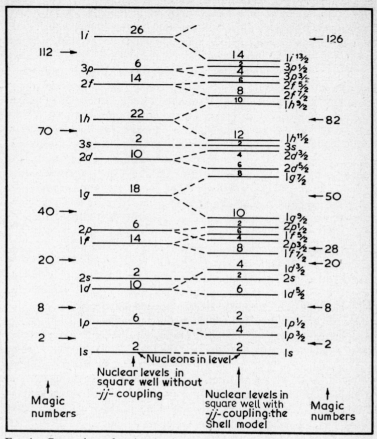

FIG. 4.   Comparison of nuclear levels calculated for a square well potential with and without -*jj*-coupling. The magic numbers predicted by the -*jj*-coupling (Shell) model are shown on the right and agree with the values found experimentally while the others do not.

made about the nuclear force, but which in the simplest case is a square well (p 10) about 40 MeV* deep and with a width equal to twice the nuclear radius. The wave equation for this system can now be solved giving the energies for all possible states. These can then be arranged in ascending order as shown in *Fig. 4*, from which it can be seen that there are wide energy gaps between certain groups of states. These gaps separate the shells, and the total number of nucleons up to the closures of the shells are the 'magic numbers', (*see also* Table 1, p 4). As in other independent particle models,

* 1 eV ≈ 1.6022 × 10^{-19} J.

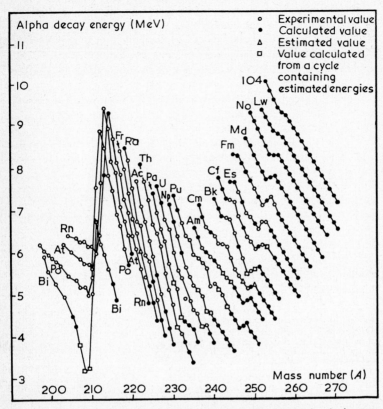

FIG. 5. α-Decay energies as a function of mass number shown for the isotopes of the elements from bismuth to kurchatovium. The sharp breaks in the curves in the region of $A = 210$ occur at the closure of the 126 neutron shell.

properties of the nucleus such as parity, angular momentum and magnetic moment depend only on the unpaired nucleons and the Shell model is very successful in predicting such static properties. The sharp breaks found in plots of functions related to these properties, *e.g.* α-disintegration energies, which occur at the shell edges lend striking support to the model (*Fig. 5*). It is, however, really only applicable to nuclei which are unexcited ('in their ground states') or have only a low level of excitation.

There are many other important models, such as the Adiabatic models which combine features of both of the foregoing types, and the Optical model which is used to interpret reactions at excitation energies of a few MeV, or more, but since these are more specialized it is left to the reader to consult the works given in the bibliography.

Stable nuclide systematics

When the elements are set out in the form of the Periodic Table many striking chemical relationships between them immediately become evident. Similarly, arrangement of the stable nuclides on a chart which has the numbers of protons and neutrons contained in them as ordinates gives some valuable information.

The first obvious fact which emerges from an examination of such a chart (*Fig. 6*) is that there is only a narrow range of proton and neutron combinations in which stable nuclei can exist. The reason is that the Pauli principle demands that the four nucleons occupying a sub-shell must consist of a pair of protons having opposite spins and a pair of neutrons also having opposite spins. When nucleons pair in this way energy (the 'pairing energy') is released, and hence the bond between them becomes stronger. The effect of this is to make the $4n$ sub-shell a particularly important unit in the construction of nuclei and the regular peaks occurring at $A = 4, 8, 12, 16$ and 24 in the inset of *Fig. 1* (p 8) clearly show the extra stability which this configuration possesses.* It might be expected, therefore, that the band of most stable nuclei would have $N = Z$, but *Fig. 6* shows that

FIG. 6. Chart showing the stable nuclides plotted as a function of their neutron (N) and proton (Z) content. The diagonal line passing through the origin is for nuclides having $Z = N$, while the backward sloping diagonal lines show nuclides having the values of A indicated on them.

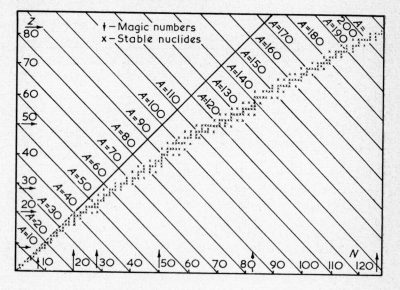

---

* This is the configuration of the α-particle.

as they become heavier, stable nuclei have relatively more neutrons than protons. This is the 'charge effect' and arises because the disrupting action of the coulomb force becomes more important as A increases, thereby requiring the addition of extra neutrons to counterbalance it.

Stable nuclides may also be classified into groups according to whether they have odd or even numbers of protons and neutrons. The significance of this classification is, once again, because of the importance of the pairing of nucleons in giving extra stability to a nucleus. This extra stability shows up quite clearly in Table 2.

Table 2. Stability of nuclei

Z	Classification N	A	No. of stable nuclides with this classification
Even	Even	Even	164
Even	Odd	Odd	55
Odd	Even	Odd	50
Odd	Odd	Even	4

Further confirmation of the extra stability granted by the pairing of nucleons can be seen if numbers of isotopes (equal Z), isotones (equal N) and isobars (equal A) are compared for odd and even values of Z, N, or A.

Another interesting feature of stable nuclide systematics which can be seen from *Fig. 6* is that in any set of isobars having a particular even value of A, the nuclides are either alternately stable and radioactive or else there is only one stable member of the set. The explanation for this is illustrated in *Figs 7, 8 & 9*. If the 'rest mass', that is, the actual mass of a bound nucleus in its ground state, is calculated by means of a mass formula (p 17) for each member of a set of even-A isobars and plotted against Z, two parabolas are obtained, the one for even-Z, even-N nuclides, lying below the one for the alternate odd-Z, odd-N nuclides because of the extra stability given by the pairing energy. The distance between the parabolas is twice the pairing energy, 2δ. Transformation from one isobar to the next is by either positive or negative β-decay (since A remains constant but Z changes), and this will take place if the rest mass of the daughter nuclide is less than that of the parent. Note that (*Fig. 9*) for the set of odd-A nuclides the two parabolas are identical and so there can only be one stable isobar. A mass parabola traces out the contour of the 'mass energy valley', one parabola being a section through a three dimensional figure whose third dimension is the range of values of A for the stable nuclides.

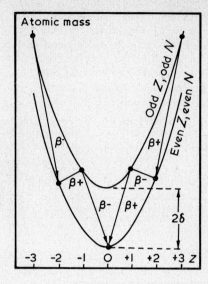

FIG. 7. Mass parabolas for a set of even-A isobars in which there are three stable nuclides. The pairing energy, δ, is sufficiently high for the isobars at $Z + 2$ and $Z - 2$ to have a lower energy than those at $Z + 1$ and $Z - 1$. (left)

FIG. 8. Mass parabolas for a set of even-A isobars having only one stable nuclide. The pairing energy is too small for the $Z + 2$ and $Z - 2$ isobars to be stable. (bottom left)

FIG. 9. Mass parabolas for a set of odd-A isobars. The parabolas for even-Z, odd N and odd Z, even N nuclides are identical (*i.e.* $\delta = 0$) and hence there can be only one stable nuclide. (bottom right)

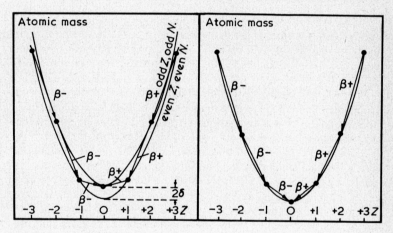

Mass and energy relationships

The most fundamental relationship between mass and energy is that of Einstein:

$$E = mc^2 \tag{6}$$

where c is the velocity of light. This equation tells us that mass and energy are completely interchangeable. The practical relationship is that one atomic mass unit is equivalent to 931 MeV.

The total binding energy of a nucleus is the amount of energy which must be put into it in order to break it down into its constituent nucleons or conversely, the amount of energy emitted when it is built up from them. It is given by the equation:

$$\Delta E = B = (Zm_p + Nm_n - m^*)\, c^2 \qquad 7$$

where m_p and m_n are the masses of the proton and neutron in free space and m^* is the mass of the assembled nucleus. In practice, m_p is usually replaced by the mass of the hydrogen atom and m^* by the atomic mass since these are the quantities actually measured. This approximation causes negligible error since the number of electrons involved is constant and the only error is due to small differences in their binding energies. The quantity within the bracket is the well known mass defect. Analogous equations can be written for any case in which nuclear transformations take place, and so in general

$$\Delta E = [(\text{mass of reactants}) - (\text{mass of products})]\, c^2 \qquad 8$$

ΔE is often called the Q-value for a nuclear reaction. Compilations of nuclear masses exist and are widely used in calculations involving nuclear reactions.

There have been many attempts to develop formulae which can be used to estimate the mass of a nucleus. These are the 'semi-empirical mass equations' all of which take the Liquid Drop model as their starting point, and all of which are essentially refinements of the original formula of von Weizäcker. Starting from equation 7 the nuclear mass can be written (neglecting the constant c)

$$m^* = Zm_p + Nm_n - B \qquad 9$$

Von Weizäcker's formula for B is:

$$B = a_1 A - a_2 A^{2/3} - a_3 Z^2 A^{-1/3} - a_4 (A - 2Z)^2/A + \delta \qquad 10$$

The five terms have the following meanings: $a_1 A$ is called the volume energy, that is, the total binding energy of all the nucleons. It is proportional to A because, as a consequence of the short range of the nuclear forces (p 8), the average binding energy per nucleon is constant.

$a_2 A^{2/3}$, the 'surface energy', results from the fact that nucleons near the surface of the nucleus have fewer neighbours with which to form bonds and so the binding of the nucleus is weakened. It corresponds to the surface tension mentioned earlier (p 11) and so it is proportional to the surface area, $A^{2/3}$.

$a_3 Z^2 A^{-1/3}$, the 'coulomb energy', allows for the disruptive effect of the protons in the nucleus which again weakens the binding. It must clearly be proportional to Z^2 and inversely proportional to the nuclear radius, itself proportional to $A^{1/3}$ (p 1).

$a_4 (A - 2Z)^2/A$, the 'asymmetry energy', is a term included to account for the fact that nuclei have an excess of neutrons over protons (p 15). The excess neutrons, $(A - 2Z)$ in number, must occupy higher energy levels than the other nucleons. They therefore have a lower binding energy than the other nucleons and so they also cause a weakening in the binding.

δ is the 'pairing energy' (p 14) and, unlike the other terms, it varies in a stepwise manner with A. It is: positive for even Z, even N; zero for odd A; negative for odd Z, odd N. The constants a_1, a_2, a_3, a_4 must be obtained by fitting known data to the equation.

Natural and induced radioactivity

Only a brief account of the essential features of this subject is given here since it is usually covered in chemistry text books.

It has been shown that nuclei in their ground states are stable if there is a balance between the nucleons giving as low a total energy (or mass) as possible. Conversely, naturally radioactive nuclei emit particles or radiations because by doing so they are transformed into a nucleus of lower mass. It may be that the transformation is hindered because the particle being emitted must get over or through a potential barrier (p 9) caused by the action of the forces within the nucleus, and this hindrance may be so strong that the nucleus is virtually stable, but nevertheless, the possibility of decay still exists. Naturally occurring radioactive nuclides decay by emission of α-particles which are ^{4}He nuclei, or β-particles which are electrons, both positive and negative, by γ-decay and by a process called electron capture. These decay modes are discussed in Chapter 3.

Apart from some 15 weakly radioactive nuclides, ranging from ^{40}K to ^{204}Pb, all the natural radioactivity on earth comes from ores containing thorium and uranium. The radioactive species found in these ores are arranged in three chains in which successive radioactive nuclides decay until a stable member is reached. Because of the establishment of 'secular equilibrium' (p 21) down the chain its overall half-life (p 20) is that of the longest lived member, and because the decays are by the emission of α-particles with an atomic mass of four or β-particles with virtually no mass, changes in A along the chain are always in multiples of four. The three chains are therefore known as the $4n$, $(4n + 2)$ and $(4n + 3)$ series. The $(4n + 1)$ series is not found in nature since it has no member with a half-life long enough for it to have survived since the formation of the elements, but it does now exist on earth because its parent, the nuclide ^{237}Np with a half-life of 2.2×10^6 years, has been made artificially. The four series are illustrated in *Figs 10, 11, 12 & 13*.

Artificial radioactivity is in principle the same as natural radioactivity. Energy is given to a stable nuclide, usually by bombarding it with particles produced by a nuclear reactor (neutrons) or accelerator (charged particles). This may raise some of its nucleons to a higher energy state, or nucleons may be added on to or knocked out of it, with the resultant formation of an unstable nucleus.

Because there are many possible routes by which a nucleus can be changed in this way, the products can decay by a greater variety of methods than naturally radioactive nuclides can. There are at least 1300 known artificially radioactive nuclides at the present time.

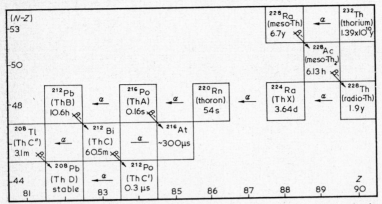

FIG. 10. The 4n (thorium) radioactive series. Since the parent of the series is ^{232}Th this is one of the natural radioactive series which exists on earth today.

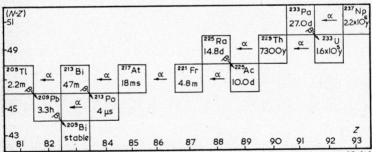

FIG. 11. The 4n + 1 (neptunium) radioactive series. This series is an artificial one since its ^{237}Np parent has now died out on earth.

The essential equations needed for calculations on radioactivity will now be given but will not be derived here.

The basic fact from which all the calculations stem is that radioactive decay is random and can therefore be treated by normal statistical methods. The decay rate of a number of atoms, N, is therefore proportional to N. Put into mathematical terms this gives the fundamental equation:

$$\frac{-\mathrm{d}N}{\mathrm{d}t} = A = \lambda N \qquad 11$$

where λ is called the disintegration constant and A is the 'activity'.

Integrating, the activity after a time t is

$$A = A_0 \mathrm{e}^{-\lambda t} \qquad 12$$

where A_0 is the original activity at $t = 0$.

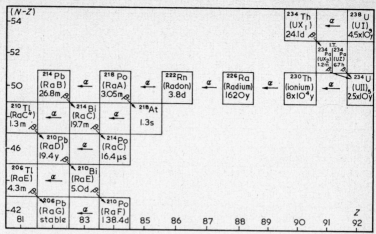

FIG. 12. The $4n + 2$ (uranium) radioactive series. Still exists on earth as a natural series.

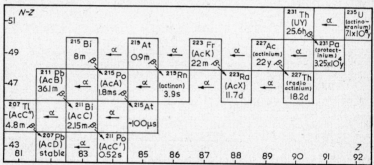

FIG. 13. The $4n + 3$ (actinium) radioactive series. Still exists on earth as a natural series.

The relationship between λ and the half-life, that is the time taken for N to be halved, is

$$t_{1/2} = \frac{\ln 2}{\lambda} \qquad\qquad 13$$

Note that the exponential nature of equation 12 means that if the logarithm of A is plotted against t a straight line of slope inversely proportional to $t_{1/2}$ is obtained.

When a daughter product is also radioactive its activity A_2 at time t is

$$A_2 = \frac{\lambda_1 \lambda_2}{\lambda_2 - \lambda_1} N_1^0 (e^{-\lambda_1 t} - e^{-\lambda_2 t}) + N_2^0 e^{-\lambda_2 t} \qquad\qquad 14$$

where N_1^0 and N_2^0 are the amounts of parent and daughter present at $t = 0$ and λ_1 and λ_2 are their respective disintegration constants.

Equation 14 reduces to

$$\frac{A_2}{A_1} = \frac{\lambda_2}{\lambda_2 - \lambda_1} \qquad\qquad 15$$

when the daughter has a shorter half-life than the parent. This is a condition called transient equilibrium and when it is established the ratio of parent to daughter activity remains constant, both decaying with the parent half-life. In the limit, if the parent half-life is so long that its decay rate is negligible, then $\lambda_2 \gg \lambda_1$ and equation 15 reduces to

$$\frac{A_2}{A_1} = 1 \qquad\qquad 16$$

that is, parent and daughter have the same activity. This is the case of secular equilibrium which is found in the natural radioactive series and which explains why natural activities with short half-lives can still exist in nature.

Finally, for the production of a radioactive substance by irradiation in a flux of particles, F, in, for example, a nuclear reactor

$$A_2 = F\sigma N_1^0 (1 - e^{-\lambda t}) \qquad\qquad 17$$

where σ is the cross section (p 35) for the production of the species A_2 from N_1^0 atoms of the parent exposed to the flux for a time t. Note that when t is very long compared with $1/\lambda$ the term inside the bracket becomes equal to one and the activity remains constant. This is a condition called saturation irradiation.

Suggestions for further reading

R. E. Marshak, 'The nuclear force', *Scient. Am.*, 1960, Mar. 99.

B. P. Nigam, 'The two nucleon interaction', *Rev. mod. Phys.*, 1963, **33**, 117.

C. M. Lederer, J. M. Hollander and I. Perlman, *Table of isotopes* (6th edn). New York: Wiley, 1967.

H. J. Mang and H. A. Weidenmüller, 'The Shell model theory of the nucleus'. *A. Rev. nucl. Sci.*, 1968, **18**, 1.

α-decay

Rutherford and Royds proved in 1909 that the α-rays emitted by some radioactive substances were nuclei of helium. It has already been shown (p 14) that the 4n configuration is especially stable and this is why radioactive nuclei emit α-particles rather than, say, deuterons. The main features of the systematics of α-decay will

FIG. 14. Modern version of the Geiger–Nuttall relationship. The logarithm of the half-life is plotted against the α-decay energy for even-Z even-N nuclides.

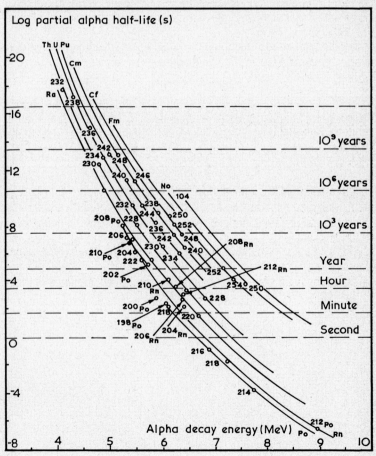

now be briefly mentioned and then the theoretical explanation of this behaviour will be given.

Geiger and Nuttall discovered in 1911 that there was a logarithmic relationship between the range (or energy) of α-particles and the half-life of the nucleus emitting them. A modern version of the relationship, which takes account of the effects of pairing energy (p 14) is shown in *Fig. 14*. This illustration is for α-emitters having an even-Z even-N configuration. α-decay energies are also strongly affected by nuclear shells (p 11) as shown in *Fig. 5* (p 13), in which there is a sharp break in the region of the 126 neutron shell. A similar break exists in the 82 neutron shell area where the rare earth α-emitters are found. The range, R, of an α-particle is related to its velocity by 'Geiger's rule',

$$R = \text{const.} \times v^3$$

Apart from the Geiger–Nuttall relationship a theory must also explain the fact that α-particles often have a kinetic energy which is less than the energy they would need to get over the coulomb barrier (p 31) before escaping from the parent nucleus. Finally, the α-particles emitted by a particular nucleus often have several sharp distinct energies which are usually quite close together (fine structure) but in a few cases a small proportion of them may have a very much higher energy (the long-range α-particles).

The main features of α-decay were satisfactorily explained by Gamow in 1928 who made use of quantum mechanics. The α-particle is formed inside the nucleus which can be represented by a potential well, and the α-particle can be pictured as rattling around inside the well, bouncing off the walls (*Fig. 15*). In classical terms the only way it can escape is by going over the top of the barrier, in which case the energy of emission will be at least equal to E_c, the coulomb

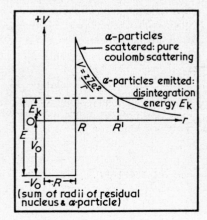

FIG. 15. Diagram to illustrate the α-decay theory. α-Particles in the potential well would require an energy at least equal to the coulomb energy to get over the potential barrier, but are able to tunnel through it and hence escape with a much lower energy, E_k.

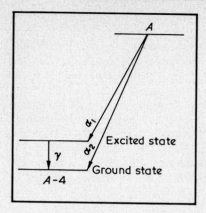

FIG. 16. Normal α-decay from the ground state of nucleus A to either ground or excited state of nucleus $A - 4$.

energy. In quantum mechanical terms, however, there is a finite probability, called the penetrability, that it will go through the barrier and if it does so its disintegration energy will be only E_k. The α-decay half-life is inversely proportional to the penetrability and this depends directly on an exponential term involving E_k. Small changes in E_k therefore cause large changes in penetrability and thus in half-life, so explaining the Geiger–Nuttall relationship.

To account for the way in which α-disintegration energies are ordered in a particular transition, it is only necessary to glance at a nuclear level diagram showing one such set of energies (*Fig. 16*). α-decay is normally from the ground state of the parent nucleus but it can be to either the ground state or an excited state of the daughter and this is why there can be groups of α-particles with slightly differing energies. The long-range α-particles are emitted from excited levels of a nucleus which is itself formed by β-decay and whose α-decay half-lives are so short that the α-particles may be emitted before the level can decay to the ground state by γ-emission. Since such levels may have energies considerably above the ground state, the α-particles they generate will have a much higher energy than those from the ground state. *Figure 17* illustrates this situation for the case of ^{212}Po.

The α-disintegration energy, E_k, is the Q-value (p 31) for the α-emission reaction. The kinetic energy of the α-particle is a little less, since some of E_k is used up in the recoil of the daughter nucleus. Thus

$$E_k = E_\alpha + E_R \qquad\qquad 18$$

This recoil energy E_R is given by

$$E_R = E_\alpha + \frac{m_\alpha}{m_D} \qquad\qquad 19$$

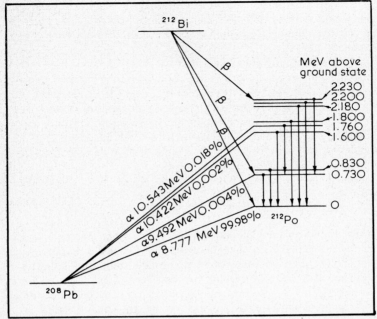

<small>FIG.</small> 17. Long range α-emitters from excited states of ^{212}Po to the ground state of ^{208}Pb.

where m_α and m_D are the masses of the α-particle and the daughter nucleus.

β-decay

The subject of β-decay is considerably more complex than that of α-decay and it can only be covered in the most general way in this monograph. Fortunately, however, from the standpoint of its chemical effects, it is only necessary to be aware of the various forms it can take and of its energetics. Readers requiring more detail, especially of the theory, are advised to consult one of the books listed in the bibliography.

The essential feature which characterizes β-decay is that in the nucleus there is no change in the value of A, but Z is either increased or decreased by one. In other words, it appears that a neutron has changed into a proton or *vice versa*. There are three possible processes by which this exchange can take place and they can be represented as follows:

$$n \rightarrow p^+ + e^- + \bar{\nu}_e \qquad \text{negatron } (\beta^-) \text{ decay} \qquad 20$$

$$p^+ \rightarrow n + e^+ + \nu_e \qquad \text{positron } (\beta^+) \text{ decay} \qquad 21$$

$$p^+ + e^- \rightarrow n + \bar{\nu}_e \qquad \text{electron capture} \qquad 22$$

2

The symbols ν_e and $\bar{\nu}_e$ represent the neutrino and anti-neutrino. They have to be included for energy balance reasons. The electrons (β-particles) emitted in any β^+ or β^- transition are not mono-energetic but cover a whole energy spectrum. However, the energy of the transition is the same whether the β-particle has a high energy or a low one, and so, except in the cases where the particle energy is at a maximum, some of the transition energy is unaccounted for. For a time it was thought that the law of conservation of energy had broken down but the need for such a grave assumption was eliminated when, in 1930, Pauli suggested that the missing energy was being carried away by an unobservable particle of essentially zero mass which he called a neutrino.

The mass and energy balance of the negatron decay process is given by

$$m_z^* = m_{z+1}^* + m_e + Q/c^2 \qquad 23$$

where the Q value is the sum of the electron and neutrino kinetic energies, together with any γ-decay energy involved in the de-excitation of the daughter nucleus to its ground state. m_z^* indicates a nuclear mass and m_e the electron rest mass. In practice, m_z^* is replaced by $(m_z - Zm_e)$ where m_z is the atomic mass and the other nuclear masses are similarly treated so that the equation becomes

$$m_z = m_{z+1} + 1 + Q/c^2 \qquad 24$$

For positron emission the mass balance equation is

$$m_z^* = m_{z-1}^* + m_e + Q/c^2 \qquad 25$$

and in atomic mass units this becomes

$$m_z = m_{z-1} + 2m_e + Q/c^2 \qquad 26$$

Note that in this case the mass difference between parent and daughter nuclide must be at least the equivalent of two electron masses ($=1.022$ MeV). The positron quickly undergoes another interaction in which it annihilates with a stray electron with the formation of a pair of annihilation γ-rays each having an energy of 0.511 MeV — a very characteristic sign of this type of β-decay.

The electron capture process, in which the nucleus absorbs an inner electron from its own electron sheath (usually a K-electron), does not require that a mass difference of 1.022 MeV exists and so this type of decay always competes with positron emission. The mass equation is

$$m_z^* + m_e = m_{z-1}^* + Q/c^2 + B/c^2 \qquad 27$$

where B is the binding energy of the absorbed electron. Using atomic masses we have

$$m_z = m_{z-1} + Q/c^2 + B/c^2 \qquad 28$$

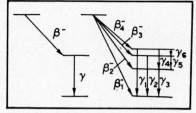

FIG. 18. Typical β-decay schemes. On the left a simple β-decay and on the right a complex one having decays to both ground and excited states of the daughter nucleus.

Half-lives for β-decay are never shorter than about 0.1 s because the interaction is 'weak'. Such long half-lives mean that an excited nucleus normally has time to revert to its ground state by γ-decay before it emits β-particles. This means that β-decay is usually from the ground state of the parent nucleus, but it may be to an excited state of the daughter. Typical decay schemes are shown in *Fig. 18*.

Pauli's concept of the neutrino paved the way for Fermi to publish in 1934 his β-decay theory in which he accounted for the shape of β-energy spectra and for the half-lives of the transitions. He compared β-decay with photon emission and he imagined the nucleus as interacting with the electron-neutrino field and in the process undergoing a transition to a final state with the creation of β-particles and neutrinos. The theory is still basically unchanged today but it has been enlarged so as to accommodate the fact that parity (p 6) is now known not to be conserved in β-decay.

γ-emission

After a nuclear process has occurred, the product nucleus may be excited, that is, it may have some excess energy in the form of rotational or vibrational energy, or some of its nucleons may be raised into orbits of higher energy than the ones they occupy when the nucleus is in its ground state. Such a nucleus will usually release this excess energy and return to the ground state by means of a γ-transition. This may involve the emission of one or more γ-rays as shown, for example, in the decay schemes pictured in *Fig. 19*, or internal conversion, or internal pair production (p 29).

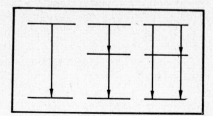

FIG. 19. Simple γ-decay schemes. Left, an excited state decays to the ground state. Centre, a simple cascade of two γ-rays *via* an intermediate level. Right, the same but with the possibility of direct decay from the highest level to the ground state.

i Classification of γ-transitions and selection rules

γ-rays are electromagnetic radiations whose frequency and energy conform to the Planck equation

$$E = h\nu \qquad\qquad 29$$

Classically electromagnetic radiation is considered to be produced by oscillating electric or magnetic charges, and this conception has survived in the classification of γ-transitions in the nucleus as electric or magnetic. The two types are distinguished only by their effect on the nucleus in which they occur: the photons from them are indistinguishable from each other. Transitions are further classified by their 'multipole order'. This is $(2)^l$ where the quantum number l is the number of units of angular momentum which are carried away by the γ-rays and so is given by

$$l = \Delta I = |I_i - I_f| \qquad\qquad 30$$

where I_i and I_f are the nuclear angular momenta of the initial and final states of the nucleus. The orders corresponding to $l = 0, 1, 2 \ldots$ are called monopole, dipole, quadrupole ... but, because of the nature of the γ-radiation the value of $l = 0$ is not allowed and so γ-rays cannot be emitted when $I_i = I_f = 0$. The γ-transition can still take place, however, by means of internal conversion, or pair production, (see below). γ-rays can be emitted when $I_i = I_f > 0$ because there is a whole range of possible I values from $(I_i - I_f)$ to $(I_i + I_f)$ but the probability falls off very sharply as l increases and so the lowest value of ΔI is normally the only one which need be considered. Electric and magnetic transitions of the same multipole order have opposite parity, $(-1)^l$ for electric and $-(-1)^l$ for magnetic.

ii Probability of γ-emission

The half-life for decay by emission of a γ-ray alone, that is after correction for internal conversion, depends on the multipolarity, on whether it is an electric or magnetic transition, on the γ-ray energy and, to a lesser extent, on the nuclear radius. Rough rules of thumb are that it is increased as follows:

1. By about 10^6 for an increase in one in multipole order for the same type of transition.

2. By about 10^2 for a magnetic over an electric transition of the same multipole order.

3. For both types of transition it increases with γ-ray energy roughly in proportion to E_γ^{2l+1}.

iii *Internal conversion and internal pair production*

When a nucleus decays by internal conversion it transfers energy directly to one of its extranuclear electrons which is then emitted with an energy

$$E = E_\gamma - B \qquad\qquad 31$$

where E_γ is the γ-transition energy and B the electron binding energy. Internal conversion always competes with γ-emission, the relative probabilities of the two modes λ_e and λ_γ, being given by the conversion coefficient

$$\alpha = \frac{\lambda_e}{\lambda_\gamma} = \frac{N_e}{N_\gamma} \qquad\qquad 32$$

where N_e and N_γ are the numbers of electrons and γ quanta emitted. The total probability of the γ-transition taking place is

$$\lambda = \lambda_\gamma + \lambda_e \qquad\qquad 33$$

and this becomes

$$\lambda = \lambda_\gamma(1 + \alpha) \qquad\qquad 34$$

by substitution in equation 32. α can have any value from 0 to ∞ and may be quoted separately for electrons from different shells, α_K, α_L . . . . Values are highest for nuclides of high Z, for γ-transitions with high l values, and for γ-transitions with low energies. Note that transitions where $I_i = I_f = 0$ are allowed by internal conversion.

The vacancy in the electron sheath is filled by an electron from a higher shell and the energy released when this happens appears either as an x-ray or as an 'Auger' electron, which is another electron from the sheath pushed out by an analogous process to internal conversion. This reorientation of the sheath continues until it has reverted to its ground state. The total numbers of x-rays or Auger electrons emitted per internal conversion electron are called the 'fluorescence yield' and the 'Auger yield' respectively.

Internal pair production is also a form of internal conversion, in which an electron–positron pair is created, but this decay method requires that the excitation energy of the nucleus is at least 1.022 MeV (twice the electron rest mass.)

iv *Nuclear isomers*

A nuclear isomer is simply a nucleus in an excited state which has a nuclear angular momentum so different from that of the ground state that γ-decay is highly inhibited and so it has a long half-life, arbitrarily usually taken as >0.1 s. The conditions necessary for the existence of isomers obtain mainly amongst certain groups of nuclei which have odd-Z or odd-N values just less than a closed shell of 50, 82 or 126 nucleons. These areas of the nuclide chart are called islands of isomerism.

v *Nuclear resonance absorption, fluorescence and the Mössbauer effect*

When a γ-ray has an energy the same within very narrow limits as that of an excited level of a nucleus, it may be absorbed so long as none of its energy is lost in recoil of the nucleus. This process is called resonance absorption and it raises the nucleus to the excitation level. If the γ-ray is now emitted, it is resonance fluorescence. Normally the emitting nucleus recoils just sufficiently to reduce the γ-ray energy below the level at which it can be absorbed by another similar nucleus. Mössbauer discovered that under some conditions, such as when the emitting and absorbing nuclei are strongly bound within crystal lattices, the γ-ray may be emitted with the full transition energy and then reabsorbed by resonance absorption. The way in which this Mössbauer effect is applied to structural chemical problems is described in Chapter 5, (p 54).

Nuclear reactions

i *The energetics*

If, in a collision between two or more atoms or particles the coulomb or nuclear forces of the nucleus are involved, a nuclear reaction has taken place. Most reactions are of the type illustrated in *Fig. 20* in which a stationary target, A, is struck by a projectile, a, giving reaction products x and X which recoil at angles θ and ϕ respectively. We can symbolize such a reaction as

$$A + a \rightarrow X + x + Q \qquad\qquad 35$$

Q is the amount of kinetic energy released or absorbed. It must be balanced by a corresponding loss or gain in the total rest mass of the system in accord with the Einstein equation ($E = mc^2$).

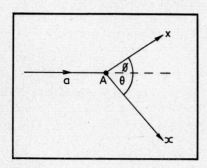

FIG. 20. A common type of nuclear reaction. Projectile a strikes the nucleus A The reaction products x and X are emitted at angles θ and ϕ to the original projectile direction.

If the Q-value is positive, meaning that energy is released and the total kinetic energy of the products is greater than that of the reactants, the reaction is called exothermic or exoergic. If negative, it is endothermic or endoergic.

The Q-value can be determined by use of the equation

$$Q = E_x(1 + m_x/m_X) - E_a(1 - m_a/m_X) - \frac{2(m_a E_a m_x E_x)^{1/2}}{m_X} \cos \theta \quad 36$$

where the E's are the kinetic energies and the m's the masses of the particles indicated by the subscripts. Corrections must be applied to this equation if the particle velocities are very high so as to allow for relativistic effects.

The equation is in 'laboratory coordinates', that is, in the frame of reference normally used and which remains stationary with reference to the laboratory. It is, however, often more convenient to use 'centre of mass coordinates' in nuclear reaction studies. This frame of reference has the centre of mass of the reacting system as its origin and so it moves in relation to the laboratory. Some of the more important mass and energy relationships between the two systems may be found in Halliday's book (*see* bibliography).

Since the Q-value is given by the change of kinetic energy, and since in the laboratory system $E_A = 0$

$$Q = E_X + E_x - E_a \quad 37$$

and this must be equal to the change in the total mass of the system, hence

$$Q = (E_X + E_x - E_a) = (m_A + m_a - m_X - m_x)\, c^2 \quad 38$$

and so Q can be determined from a knowledge of the rest masses or conversely an unknown mass can be determined if the othe three are known and Q is measured.

Endothermic reactions will not occur until a minimum threshold energy is applied to the system, that is, until the kinetic energy of the particle a reaches this value. It is given by

$$E_T = -Q \left[\frac{m_a + m_A}{m_A} \right] \quad 39$$

that is, it is the Q value corrected for the recoil energy of the target nucleus A.

Finally, an additional complication arises if a is a charged particle because before it can get near enough to the nucleus to react with the short range nuclear forces, it must overcome the electrostatic repulsion from the protons, the coulomb barrier. Classically this is equal to

$$E_c = \frac{Z_a Z_A e^2}{r_a + r} \quad 40$$

Z_ae and Z_Ae being the particle and nuclear charges and r_a and r_A their radii.   Neutrons do not, of course, meet such a barrier.

ii The mechanism by which nuclear reactions take place

Broadly speaking, low energy reactions, that is, those where the kinetic energy of the projectile is below about 50 MeV, take place through the formation of a compound nucleus as proposed by N. Bohr in 1936.   Above this energy direct interactions become more and more important but there is no precise dividing line separating the two types.

In a compound nucleus reaction the target and projectile fuse together to form a nucleus which has an excitation energy

$$E_x = E_B + E_a \cdot m_A/(m_A + m_a) \qquad 41$$

where E_B is the binding energy of the projectile in the compound nucleus in its ground state and E_a its kinetic energy in the laboratory system.   The lifetime of the compound nucleus, which is of the order of 10^{-14} s, is very long when compared to the time taken by a particle to cross it of about 10^{-21} s, because the excitation energy is rapidly spread over all the nucleons and so it takes time before enough of it can be concentrated on one or more nucleons to cause them to be emitted.   Because of this, the way in which the compound nucleus decays depends only on its own properties and excitation energy and not on those of the reactants.   Various modes of decay may compete with each other, the probability of each being expressed by its 'level width' Γ.   The total decay probability is then given by the sum of these partial level widths.

$$\Gamma = \Gamma_\alpha + \Gamma_{\text{fission}} + \Gamma_\gamma + \ldots \qquad 42$$

Direct interactions are quite different.   The oncoming particle strikes a nucleon which may be immediately ejected or may recoil inside the nucleus.   Other nucleons may be struck, either by the original projectile or by the recoiling nucleon and so a number of 'knock-on' particles or even groups of particles may be rapidly emitted with high energies.   This is the 'cascade' and it leaves the nucleus with excitation energy which it then loses by 'boiling off' more particles, by fissioning, or by emitting γ-rays.   This is the evaporation stage. The mathematical treatment of direct interactions is difficult and the reader is referred to the bibliography for further information.

iii The main types of nuclear reactions

The main types of reaction will now be discussed but the grouping is made for the sake of convenience and should not be regarded as being rigid: the various groups may merge into each other.   Table 3 roughly summarizes the main reactions and shows the order in which they become possible.

Table 3. Summary of reactions for nuclei of intermediate mass ($30 < A < 90$) and for heavy nuclei ($A > 90$)

Incident particle / Particle energy	Intermediate nuclei				Heavy nuclei			
	n	p	α	d	n	p	α	d
Low 0–1 keV	n(el), γ, (res)				γ, n(el), (res)			
Intermediate 1–500 keV	n(el), γ, (res)	n, γ, α, (res)	n, γ, p, (res)	p, n	n(el), (res), γ, (res)			
High 0.5–10 MeV	n(el), n(inel), p, α, (res. for lower energies)	n, p(inel), α, (res. for lower energies)	n, p, α(inel), (res. for lower energies)	p, n, pn, 2n	n(el), n(inel), p, γ	n, p(inel), γ	n, p, γ	p, n, pn, 2n
Very high 10–50 MeV	2n, n(inel), n(el), p, np, 2p, α, three or more particles	2n, n, p(inel), np, 2p, α, three or more particles	2n, n, p, np, 2p, α(inel), three or more particles	p, 2n, pn, 3n, d(inel), tritons, three or more particles	2n, n(inel), n(el), p, pn, 2p, α, three or more particles	2n, n, p(inel), np, 2p, α, three or more particles	2n, n, p, np, 2p, α(inel), three or more particles	p, 2n, np, 3n, d(inel), tritons, three or more particles

Abbreviations: el = elastic; inel = inelastic; res = resonances. The abbreviation (res) refers to all reactions listed in the group. The symbols indicate the particle emerging from the reaction, while the order of the symbols in a particular group corresponds roughly to the yields of the corresponding reactions. Fission and elastic scattering of charged particles are omitted. Nuclei with $A < 30$ are not shown because it is not possible to classify their reactions in any simple manner. (Reprinted, by permission, from: J. M. Blatt and V. F. Weisskopf, *Theoretical nuclear physics*, New York: Wiley, 1952.)

Elastic scattering. The products of the reaction are the same as the reactants and there is no change in total kinetic energy so the *Q*-value is zero. It is as if the projectile had simply bounced off the target nucleus. There are three main types of elastic scattering. In Rutherford or coulomb scattering, the interaction is with the coulomb force of the protons. The second kind is variously known as potential scattering, hard sphere scattering or shape elastic scattering and is an interaction at the nuclear surface with the nuclear force itself. Finally, there is elastic resonance scattering, or compound elastic scattering, in which a compound nucleus is formed but decays by emitting an identical particle with the same kinetic energy as the incoming projectile.

Inelastic scattering. This is similar to compound elastic scattering but the emitted particle has a lower kinetic energy than the projectile, which leaves the target nucleus in an excited state.

Transmutation reactions. These are reactions in which the products are different from the reactants, but for the sake of convenience some reactions to which this definition does apply are given separate names. These are the ones mentioned in the following paragraphs. At very low energies the only reaction which is important (apart from fission) is one in which a neutron is absorbed and a γ-ray emitted. The shorthand way of indicating a reaction of this type is

$$^{A}X(n, \gamma)^{A+1}X$$

The reactions are exothermic and are often referred to as radiative capture reactions. They are very important because low energy neutrons are plentiful in thermal nuclear reactors, and, hence a high proportion of commercially available isotopes are made by means of these reactions.

As the excitation energy of the compound nucleus is increased, probably by raising the kinetic energy of the projectile, other reactions become energetically possible and competition between them sets in as shown in Table 3.

Photonuclear reactions. Reactions such as (γ, n), or (γ, f) initiated by photons are called photonuclear reactions. They are always endothermic.

Stripping and pickup reactions. 'Stripping' is a direct interaction process in which the projectile, which is frequently a deuteron because its two nucleons are not very strongly bound together, leaves behind one of its nucleons in the target nucleus. The inverse reaction in which one is added to the projectile is 'pickup'.

Spallation. This name is given to a transmutation in which the excitation energy is sufficiently high so that any one of a series of products may be formed. It often competes with fission.

Fission. The breakup of the target nucleus into two large pieces plus a number of smaller ones after a compound nucleus has been formed. This is such an important transmutation reaction in practice that it is described separately.

Fragmentation. A direct interaction very high energy process in which the target nucleus breaks rapidly into several sizeable fragments.

iv *Cross-sections*

If it is established that a certain reaction is energetically possible we then want to know what is the probability that it will take place. This is expressed by the cross-section, which may be thought of as the area of the target nucleus within which the reaction will certainly occur if the projectile strikes it. This area may be equal to the geometrical area of the target but is usually larger or smaller, often by many orders of magnitude. It is measured in units of cm^2 or barns where one barn is equal to 10^{-24} cm^2.

The theoretical prediction of cross-sections is very complex and the method used depends very much on the energies involved, the nature of the target and projectile, the type of reaction concerned, *etc.* This will not be discussed here. If, however, the number of nuclear reactions taking place per second, R, can be measured for a target containing N nuclei per cubic centimetre and having a thickness t, the cross-section is

$$\sigma = \frac{R}{nvNt} \qquad\qquad 43$$

where nv is the flux of projectiles per square centimetre per second.

Nuclear fission

i *General description of the fission process*

By the definition given on p 34 fission is a transmutation reaction. It has been described under a separate heading, partly because of its extreme importance in both economic and military terms, and partly because it is an exceedingly complex nuclear reaction with important characteristics of its own.

When a nucleus, whether a compound nucleus (p 32) formed by the fusing of a bombarding particle and a heavy nucleus, or a nucleus in its ground state, breaks up into two pieces of comparable size together with a number of light particles, the reaction is called fission; in the first case 'induced fission', in the second 'spontaneous fission'.

Fission is a possible mode of decay for any nucleus whose breakup results in products having a lower total rest mass (*see* p 15), that is, for all nuclei above iron in the Periodic Table. The reason why it occurs only with difficulty for all but the very heavy nuclides is

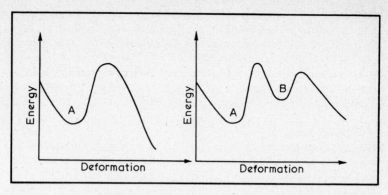

Fig. 21. Fission potential barriers in which the energy of the nucleus is plotted as a function of its shape (deformation). Left, a simple barrier: the ground state of the nucleus is at a deformation corresponding to the minimum A. Right, a double-humped barrier; the ground state is again at A but if the nucleus is further stretched out it may remain for a time at the deformation of the secondary minimum, B, corresponding to a spontaneous fission isomer.

that there is a potential barrier which must either be surmounted (induced fission) or penetrated (spontaneous fission) before the lower mass state can be reached. This 'fission barrier', illustrated in *Fig. 21*, develops as the nucleus is stretched out. Energy must be put in to stretch it out against the pull of the nuclear force, and the potential energy rises. Beyond a certain point where the potential is highest, usually called the saddle point, the nuclear force weakens rapidly and the potential drops again until fission occurs at the scission point. The two positively charged parts of the nucleus, called fission fragments, fly apart and energy is given out as their kinetic energy and as the kinetic energy of the light 'prompt particles' (neutrons, protons and alpha-particles). Further energy is contained in the fragments as their excitation energy, and they lose this by emitting neutrons and γ-rays until they are in their ground state. Finally, the fragments, which have re-formed their electron sheaths and are now known as fission products, decay over a time scale ranging from about 1/10 s to many years, by emitting β-particles and γ-rays and, in a few cases, 'delayed neutrons'. This whole sequence of events is shown in *Fig. 22*.

As the nucleus stretches out the change in potential, which is the result of a complicated inter-play of the various nucleon levels and the transfer of nucleons between them, may either take the simple form shown on the left of *Fig. 21* or the more complicated one on the right, in which the barrier has two humps with a secondary minimum between them. It is then possible for the highly distorted nucleus to exist at the level of this secondary minimum for a while before

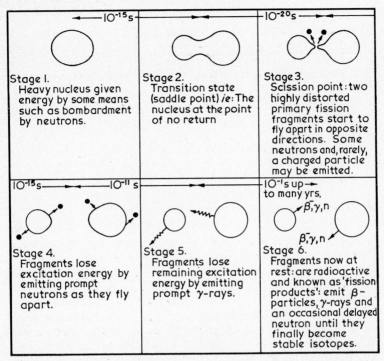

FIG. 22.   The sequence of events in fission of a single nucleus.   (Reprinted, by permission, from *Nuclear fission*, J. G. Cuninghame.   Chicago: Encyclopaedia Britannica, 1973.)

decaying by spontaneous fission.   Such nuclear states are called spontaneous fission isomers.

It is obvious that the shape of the barrier is very important, since energy at least equal to its height must be given to the nucleus if fission is to be induced in it.   Spontaneous fission is similar to α-decay (p 22) and the fission half-life depends sharply on the shape of the barrier, particularly its width.   Fission barrier shapes depend on the number and arrangement of nucleons in the nucleus, but in general they are higher near to magic numbers of nucleons and gradually become lower as we move to heavier and heavier nuclei. This is why fission becomes so much more important towards the end of the Periodic Table.

ii *Energetics*

The amount of energy released in fission depends on the fissioning nucleus, but is of the order of 200 MeV.   In the case of fission of ^{235}U by thermal neutrons, for example, the total release is 219 MeV, made up as shown in Table 4.

Table 4.   Energy release in fission of ^{235}U by thermal neutrons

		MeV
Kinetic energy of fission fragments	⎫	166
Kinetic energy of prompt neutrons	⎪ The 'energy	5
Binding energy of prompt neutrons	⎬ release in	12
Energy carried by prompt γ-rays	⎪ fission'	8
Energy of β-particles	⎭	8
Energy of anti-neutrinos accompanying the β-particles		12
Energy of delayed γ-rays		8
		—
		219

Note that the anti-neutrino energy cannot be utilized since these particles do not take part in reactions with matter to any significant extent.

iii *The fission fragments*

The fission fragments, that is, the two largest pieces of the fissioning nucleus, are known as fission products once they have emitted their prompt neutrons and γ-rays and have restored their electron sheath. They are, of course, radioactive isotopes of elements from the middle part of the Periodic Table.   Fission is not equally probable into all possible pairs of fission products, and they have a distribution with respect to their mass whose shape depends on the fissioning compound nucleus and on its excitation energy.   Thermal neutron fission of heavy nuclei such as ^{235}U and ^{239}Pu is asymmetric, *i.e.*, the nucleus preferentially splits into two unequal parts, having most probable masses of about 100 and 140.   Other types of fission may have just one broad peak, or even three equal ones.   Some examples of mass distributions are shown in *Fig. 23*, which also gives an example of the distribution of prompt fission neutrons as a function of fission fragment mass.

iv *Theories of fission*

Because of the complex character of the nuclear forces and because so many nucleons are involved, a fundamental mathematical treatment of the fission process cannot be attempted at the present time and models are therefore used (p 11).   The first to be used in this way was the Liquid Drop model (p 11).   As originally formulated it predicted that the mass distributions of the fission products in low energy fission of heavy nuclei would be symmetrical, and although modern versions of the model have got over this problem, they have introduced further difficulties into the calculations.

The other two main fission models are the Adiabatic model and the Statistical model.   The former is quite useful for describing the first part of the fission process as the nucleus stretches out to the saddle point, and the latter has been reasonably successful at the

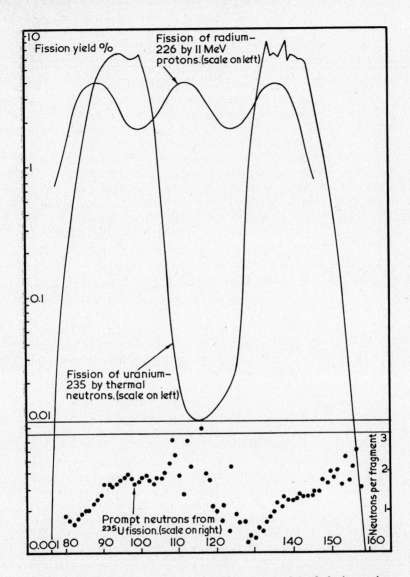

FIG. 23. Two examples of the way in which the mass of the fissioning nucleus is distributed after fission. Curves show the probability that fragments will have a particular mass number expressed as a percentage of the number of fissions taking place. Below, the dots show how many prompt fission neutrons are emitted accompanying fragments of a particular mass for the case of ^{235}U thermal neutron fission. (Reprinted, by permission, from *Nuclear fission*, J. G. Cuninghame. Chicago: Encyclopaedia Britannica, 1973.)

other end and can now predict mass distributions quite well. Readers wanting a clear and detailed comparison of the various models should consult the monograph by Wilets referred to in the bibliography.

v *Chain reactions*

When a ^{235}U nucleus experiences low energy fission about 2.5 prompt neutrons are released on average. Some of these may cause other uranium nuclei to fission and so a self-sustaining chain reaction begins. If the average number of neutrons from each fission which cause another fission is much greater than one the chain multiplies very rapidly and a nuclear explosion ensues, while if it is below one the chain dies out. A self-sustaining chain reaction cannot occur if the mass of fissile isotope is below a certain critical mass because too many neutrons are lost to the outside. In a nuclear reactor the 'multiplication factor' is kept very close to one by the adjustment of 'control rods' of a neutron absorbing substance in the reactor core.

vi *Utilization of fission*

Everybody knows that the first open use made of nuclear fission was in the atomic bombs dropped on Hiroshima and Nagasaki in 1945, but in fact Fermi had built the world's first nuclear reactor in which a controlled chain reaction was achieved in Chicago in 1942. Since those days, while military uses have retained their importance, the heat from nuclear reactors has been increasingly used for the more fruitful purpose of generating steam to provide electricity. There is no doubt that with the rapid using up of fossil fuel resources, by the year 2000 fission reactors will dominate the electricity supply industry.

Nuclear reactors have also been used in ship propulsion but they have not yet become economically viable for this purpose. Nuclear submarines are extensively used by the British, Russian and American navies, however, but economic operation is of secondary importance in this case.

The other major use of fission arises out of the fact that many essential radioactive isotopes needed for medical and industrial purposes can be prepared by means of the (n, γ) reaction (p 34). This reaction often has high cross-sections (p 35) at thermal energies so the isotopes can easily be made by irradiation of the target substance in a reactor.

Suggestions for further reading

I. Halpern, 'Nuclear fission', *A. Rev. nucl. Sci.*, 1959, **9**, 320.
'Nuclear fission' by J. R. Huizenga and R. Vandenbosch, in *Nuclear reactions* vol. II (P. M. Endt and P. B. Smith eds). Amsterdam: North-Holland, 1962.

'Fission phenomena' by E. K. Hyde in *Nuclear properties of the heavy element*, vol. III. Englewood Cliffs, N.J.: Prentice-Hall, 1964.

L. Wilets, *Theories of nuclear fission*. London: OUP, 1964.

R. B. Leachman, 'Nuclear fission', *Scient. Am.*, 1965, Feb., 49.

J. S. Fraser and J. C. D. Milton, 'Nuclear fission', *A. Rev. nucl. Sci.*, 1966, **16**, 379.

'Nuclear fission' by J. E. Gindler and J. R. Huizenga in *Nuclear Chemistry*, vol. II (L. Yaffe ed.). New York: Academic, 1969.

'Nuclear fission' by J. G. Cuninghame in *Encyclopaedia Britannica*, 1973.

4. Interaction of Radiation with Matter

In this short chapter the ways in which radiations interact with bulk matter will be summarized. While some nuclear interactions do take place, most of the energy of a particle or ray is absorbed in matter by reaction with the atomic electrons. Neutrons, whose energy is absorbed almost exclusively by nuclear reactions are an exception however.

Heavy charged particles

Under this heading are included all the charged particles except electrons. The α-particle, which has been the most intensively studied, serves as a suitable model for them all.

At the beginning of its track through matter, an α-particle loses energy by elastic collisions with the outer electrons of atoms in the medium. These collisions result in primary ionization of the struck atoms or in their excitation, and the electrons recoiling from these collisions may produce further secondary ionization or 'δ-rays'. The number of ion-pairs produced per mm of track in air at 15 °C and 760 mm pressure is called the specific ionization. When the α-particle has been slowed down to velocities comparable with the orbital velocity of helium K-electrons, it loses the remainder of its energy by exchanging electrons with those in atoms of the medium. Nuclear reactions occur too, but they are a negligible cause of energy loss.

Because two totally different processes are involved in bringing an α-particle to rest, its 'range', that is the depth of medium it can penetrate, is not related to its energy in any simple way. The easiest way of relating them is to use range–energy curves experimentally determined for a particular medium, some examples of which are shown in *Fig. 24*. A related concept is the 'stopping power' of an absorber medium. This is the rate of energy loss in MeV per millimetre of track in the medium.

Very heavy charged particles, such as the charged carbon and oxygen ions produced by many modern accelerating machines, and fission fragments, also lose most of their energy by electronic collisions. Towards the end of their range, however, nuclear elastic scattering becomes a much more important source of energy loss than it is with the lighter ions.

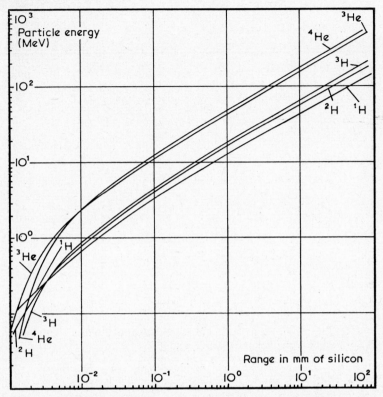

FIG. 24.   Range–energy curves for charged particles in silicon.

Electrons

Electrons at fairly low energies lose most of their energy by collision
with the outer atomic electrons in the same way as charged particles
do, but because the masses involved are the same, up to half the
energy can be lost in a single collision.   At higher energies inelastic
nuclear scattering takes place with production of *bremsstrahlung*
(p 45) while at still higher energies it becomes necessary to allow for
relativistic effects.   Elastic scattering with both nuclei and electrons
is not a major source of energy loss but does result in large deflections
and so electron tracks in an absorber are very crooked.   Note that
the negatrons and positrons behave very similarly except that towards
the end of its track a positron will annihilate with an electron with the
emission of γ-rays (p 32).   Range–energy relationships for electrons
are even more complicated than for heavy charged particles, and
once again it is simpler to use practical curves such as the one in
Fig. 25.

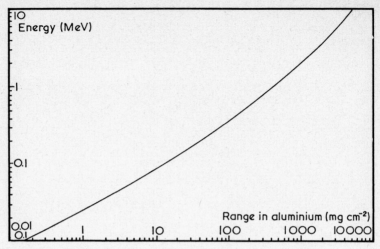

F<small>IG</small>. 25.   Range–energy curve for electrons in aluminium.

γ-rays

There are three important ways by which γ-rays lose energy.   The first is called the photo-electric effect.   It is similar to internal conversion (p 29) in that the γ-ray is absorbed by the atom and an electron is emitted, having kinetic energy

$$E_k = E_\gamma - B \qquad\qquad 44$$

where B is the electron binding energy.   Rearrangement of the electron sheath occurs with emission of x-rays. K-electrons are emitted preferentially so long as E_γ exceeds their binding energy and there is a sharp fall off of the photo-electric effect at a value of E_γ equal to this.   This is termed the 'K-edge'.   The cross-section for photo-electric absorption is approximately related to the atomic number of the absorber by

$$\sigma_{PE} \propto Z^5 \qquad\qquad 45$$

and to the γ-ray energy by

$$\sigma_{PE} \propto 1/E_\gamma^{7/2} \qquad\qquad 46$$

In the Compton effect the γ-ray loses some energy to an electron and is deflected in the process.   The amount of energy lost is not constant as it is in the photo-electric effect but is continuously variable up to a limit (the Compton edge).   Rough experimental rules for the Compton effect cross section are

$$\sigma_c \propto Z \qquad\qquad 47$$

$$\sigma_c \propto 1/E_\gamma \qquad\qquad 48$$

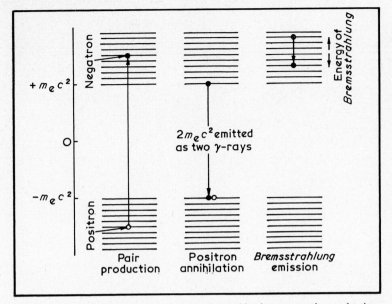

FIG. 26. Graphic illustration of the relationship between pair production, positron annihilation, and *bremsstrahlung* emission. In the first a γ-ray is absorbed and the energy raises an electron to a positive level, leaving a hole which is observed as positron, *i.e.* fills up the hole. In the second, an electron annihilates with a positron, *i.e.* fills up the hole. In the third, an electron drops to a lower positive level and the energy difference is emitted as a photon (*bremsstrahlung*). The lowest positive and negative electron states are separated by $2m_e c^2$ *i.e.* by twice the electron rest mass.

The third important way in which γ-rays may lose energy is called pair production. It is related to positron annihilation and to *bremsstrahlung* emission as *Fig. 26* shows, and is a reaction with the nucleus. The absorption of the γ-ray causes an electron to be raised from a negative to a positive energy level, leaving a 'hole' which is observed as a positron. The γ-ray must have an energy at least equal to two electron rest masses ($=1.022$ MeV) before the process can occur. A rough rule for the cross-section is

$$\sigma_{PP} \propto Z^2 \qquad\qquad 49$$

The positron finally annihilates with an electron and two 0.511 MeV γ-rays are emitted. *Bremsstrahlung* are γ-rays resulting from the transfer of an electron from a positive energy level to a lower one, the photon energy being the energy difference between the levels. An example showing how the three main γ-absorption processes relate to each other for the case of a lead absorber is shown in *Fig. 27*.

Finally, there are several other minor ways in which γ-rays can lose energy. One of these which is of importance in nuclear physics

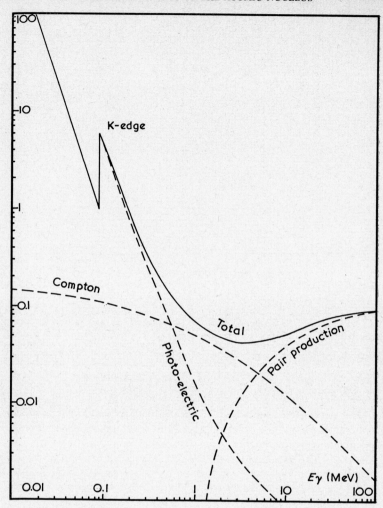

FIG. 27. Absorption of γ-rays by lead. Solid line shows the total absorption of the primary radiation in arbitrary units as a function of γ-ray energy. The dotted lines show the contributions of the three main absorptive processes.

is the (γ, n) or 'radiative capture' reaction, in which γ-rays are absorbed by a nucleus and neutrons are emitted. Another, which has become important because of its connection with the Mössbauer effect (p 30), is 'nuclear resonance absorption'.

Neutrons
Neutrons, being uncharged, lose their energy almost exclusively by means of nuclear reactions. Of these, nuclear inelastic scattering

is the most significant at the higher energies, but elastic scattering becomes important as the energy falls. Transmutation reactions can also absorb neutron energy, and once the neutrons have been slowed down to thermal energies, neutron capture usually takes place. It is also possible for the neutrons to decay by β-emission.

Suggestions for further reading

Experimental nuclear physics, vol. 1 (L. Segrè ed.), p 166. New York: Wiley, 1953.

L. C. Northcliffe and G. Schelling, 'Range and stopping power tables for heavy ions'. *Nuclear data tables*, 1970, **7**, 233.

E. A. Uehling, 'Penetration of heavy charged particles in matter'. *A. Rev. nucl. Sci.*, 1954, **4**, 315.

Bibliography

Books covering more or less the whole of the subject matter of Part 1.

D. Halliday, *Introductory nuclear physics* (2nd edn). New York: Wiley, 1955.

J. G. Cuninghame, *Introduction to the atomic nucleus.* Amsterdam: Elsevier, 1964.

G. Friedlander, J. W. Kennedy and J. M. Miller, *Nuclear and radiochemistry* (2nd edn). New York: Wiley, 1966.

B. G. Harvey, *Introduction to nuclear physics and chemistry* (2nd edn). New York: Prentice-Hall, 1969.

P. Marmier and E. Sheldon, *Physics of nuclei and particles.* New York: Academic, 1969.

W. M. Gibson, *Nuclear reactions.* Harmondsworth: Penguin, 1971.

J. M. Reid, *The atomic nucleus.* Harmondsworth: Penguin, 1972.

Part 2. The Nucleus Applied to Chemistry

5. Structure Analysis

Introduction

In the most general terms atomic structure analysis consists of attempting to find the relative positions of the atoms forming the material under investigation, the distribution of the electrons round them, and the forces between them. The principle behind all of the methods described in this chapter is that some effect is measured whose value is sensitive to small alterations in the structure. Since this monograph is concerned with the intrusion of the nucleus into chemistry, the effects to be examined are mostly nuclear ones. For the sake of completeness, however, x-ray diffraction and x-ray spectroscopy are also discussed but, even though the former is probably the most important tool of the structural chemist today, only in outline. These subjects are, of course, normally covered in chemistry books.

X-ray diffraction

X-ray diffraction is completely analogous to the diffraction of light, but because x-rays have a much shorter wavelength they are diffracted by much finer gratings. Their wavelengths are, in fact, comparable with inter-atomic distances in a crystal lattice which is therefore capable of diffracting them. When this happens the lattice structure can be studied by examining the directions and intensities of the diffracted beams because these parameters depend on the nature and position of the atoms in the crystal.

The principles on which the method is based are quite simple. When the x-ray beam strikes the crystal the rays are scattered by the atoms in the lattice. The scattered rays from all the atoms in one single plane form a wave train. If the difference in path length between this wave train and that from the one scattered from the next parallel plane of the lattice is an exact number of wavelengths the result will be an intensification of the wave. If it is not, the wave intensity will be reduced. The diffraction pattern produced depends, therefore, on the distance between successive lattice planes. The quantitative relationship between the various parameters is known as Bragg's Law after its discoverer, L. Bragg. It states that:

$$n\lambda = 2d \sin \theta \qquad\qquad 50$$

where λ is the x-ray wavelength, d is the distance between successive lattice planes, and θ is the angle between the beam and the diffracting plane. n, the order of the diffraction, is an integer showing how many wavelengths there are in the path length difference. As written above, the law only allows for diffraction in one direction, but since a crystal lattice is three-dimensional there must be three order integers, usually represented by the letters h, k, and l.

Single crystals of the materials under investigation are normally employed and give by far the most information, but in some instances it is possible to use a micro-crystalline powder, although this gives much more limited information and is successful only in simple cases. The powder method is often used where two similar structures are to be compared, and must be employed, of course, if crystals cannot be obtained, for example when alloys are examined. It is used extensively for analytical purposes however.

The very first crystal structure established by x-ray diffraction was that of sodium chloride, by W. H. Bragg, in 1913. Since that time the method has been applied to more and more complex crystals, and with the advent of digital computer analysis, which has removed much of the drudgery from the interpretation of the results, large biochemical molecules such as proteins and enzymes are now being successfully analysed. An example of a typical x-ray photograph is given in *Fig. 28*.

Neutron diffraction

The principles behind neutron diffraction are exactly the same as for x-ray diffraction, but there is one major difference in the mechanism by which the diffraction takes place: x-rays are scattered by the electron sheath of the atom, while neutrons are scattered by the nucleus. Two main consequences spring from this fact. First of all, the scattering amplitude of x-rays depends on the number of electrons in the atom and so it increases with the atomic number and this makes the correct localization of the light atoms difficult if there are heavy atoms present in the lattice. Neutron scattering amplitudes, on the other hand, do not vary with the atomic number in such a simple way and the variation is, in any case, not very large (*see* below). Secondly, there will be interference between x-rays scattered from different parts of an atom because x-ray wavelengths (10^{-8}–10^{-9} cm) are comparable with its size. The degree of interference depends upon the Bragg angle of the beam and this interaction markedly affects the scattering amplitude. The wavelength of a beam of thermal neutrons, however, is large ($\sim 10^{-8}$ cm) compared with the nucleus from which the neutrons are scattered (radius $\sim 10^{-13}$ cm) and so there is no such effect for them. Scientifically neutron diffraction is probably a better tool for structural analysis than is x-ray diffraction, but its much greater expense, together with the fact that suitable reactor neutron beams are not widely available, militates sharply against it.

FIG. 28. Zero level Weissenberg x-ray photograph of $UO_2(NO_3)_2$ $6H_2O$ about the b_0 axis. The positions and intensities of the reflections can be used to determine the unit cell dimensions and the structure of the compound respectively (but many more data are needed for a reliable determination of the structure). α_1 and α_2 reflections are resolved at the top and bottom of the film. (Recorded by P. T. Moseley of AERE, Harwell.)

It has been mentioned that the variation in neutron scattering amplitude with atomic number is irregular. This is because the total elastic neutron scattering cross-section is made up of two parts (p 34), potential scattering in which the neutron is deflected by the nuclear potential at the nuclear surface, and compound elastic scattering in which a compound nucleus is formed between the neutron and the nucleus and then decays by emitting a neutron with the same kinetic energy as the incident neutron. The potential scattering part of the cross-section is proportional to the nuclear radius and therefore to $A^{1/3}$, while the magnitude of the compound elastic scattering part depends on the closeness of the match between the neutron energy and that of the virtual level of the compound nucleus which is formed. It is, in fact, a resonance phenomenon. The above picture applies to the case where the scattering nucleus has zero nuclear angular momentum, I. If it has a value of $I > 0$, the compound nucleus can have either of the values $(I + \frac{1}{2})$ or $(I - \frac{1}{2})$, and this leads to further irregularities in the behaviour of the compound elastic part of the scattering cross-section. It also leads to the production of incoherent scattering (scattering in which the radiation emitted is different both in wavelength and phase from the original radiation) which gives rise to a background of undiffracted neutrons.

One final type of neutron diffraction should be mentioned. When the scattering atoms have an electronic magnetic moment this may couple with the neutron magnetic moment and diffraction of the neutrons by the electron sheath of the atom may occur. Such diffraction is subject to the same interference effects as are x-rays. In terms of cross-section it is small compared to the nuclear diffraction but it has proved capable of yielding valuable information about the magnetic structure of crystals.

X-ray spectroscopy

While diffraction methods are useful for providing information as to the position of atoms in a lattice, x-ray spectroscopy is concerned with the energy levels of atoms and with the behaviour of their electrons.

X-rays arise because an atomic electron is removed from its place in the electronic sheath of an atom. This may occur as a result of radioactive decay of the nucleus, *e.g.* in internal conversion (p 29), or electron capture (p 26), or it may be due to transfer of all or part of the energy of a photon or electron which strikes the atom from outside. However it comes about, a vacancy is created in the electronic sheath which is then filled by an electron dropping in to it from a higher energy orbit, with the resultant emission of an x-ray photon with an energy equal to the energy difference between the two orbits. This energy depends on which two orbits are concerned in the electron transition and on the atomic number of the atom in question. For any particular transition the energy is slightly altered by changes in the physical or chemical state of the solid of which the atom is part, and hence precise measurement of the energy can give information about such changes. The energy (E) and frequency (v) of the x-rays are related by the Planck equation

$$E = hv \qquad\qquad 51$$

where h is Planck's constant; and the frequency for any particular transition is related to the atomic number (Z) by Moseley's equation,

$$\sqrt{v} = k_1(Z - k_2) \qquad\qquad 52$$

where k_1 and k_2 are constants.

In emission x-ray spectroscopy the wavelengths of the various lines in the spectrum are accurately measured after the atoms have been excited by bombarding them with electrons or with x-rays. In absorption spectroscopy the intensity of a beam of x-rays passing through the substance is studied as a function of the x-ray wavelength. At short wavelengths absorption is mainly due to transfer of energy to K-electrons. As the wavelength gets longer, (*i.e.* as energy drops) the absorption increases until at some point the x-ray energy is just insufficient to remove a K-electron from its orbit, and so the absorption falls off abruptly (*Fig. 29*) to rise again as transfer to an L-electron takes over. The energies of these 'absorption edges' depend on the same factors as do the energies of the emission lines, but they are larger because they are the energies required to remove electrons right out of the atom. The object of both emission and absorption spectroscopy in structure analysis is to correlate the small energy changes observed with alterations to the electronic structure of the substance concerned. Direct analytical applications are discussed later (p 80).

There are many types of emission x-ray spectrometer. The most precise of them utilize the diffraction of the x-rays from a crystal or a ruled grating. If the Bragg angle (p 49) of the emergent beam is measured, its wavelength can be calculated from the Bragg relation. The ray is detected by either some sort of ionization device (ion-chamber, proportional counter, Geiger counter, semiconductor detector) if the intensity is required, or by photographic emulsion if it is not. *Figure 30* shows the experimental arrangement in diagrammatic form. Lithium-drifted semiconductors are also being increasingly used for x-ray spectroscopy. They have the advantage of being much simpler to use than grating spectrometers, have a much higher efficiency and are being rapidly improved. At the

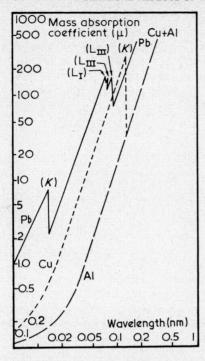

FIG. 29. The absorption of x-rays by lead, copper and aluminium. Absorption edges of the K and L electrons in lead and of the K electrons in copper can clearly be seen. (Reprinted, by permission, from H. A. Liebhafsky, *Ann. N. Y. Acad. Sci.*, 1960, **53**, 997.)

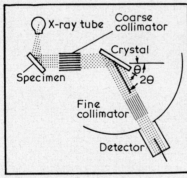

FIG. 30. One type of experimental arrangement used for x-ray spectroscopy. The x-rays are diffracted from a plane crystal, the diffraction angle depending on wavelength (energy) in accordance with the Bragg relation. (Reprinted, by permission, from *X-ray spectroscopy*, H. Friedman in *Advances in spectroscopy*, vol. 11. H. W. Thomson (ed.). New York: Interscience, 1961.)

present time they are capable of an energy resolution which is, in round terms, only about a factor of two worse than grating spectrometers.

Mössbauer spectroscopy

i *Theoretical discussion*

The γ-rays emitted from a free atom in any particular γ-transition have an energy which differs from the actual transition energy in two ways. In the first place the line width of the γ-rays, Γ, is much greater than would be expected from application of the Uncertainty Principle. According to the latter, the energy and lifetime of a level are related by the equation

$$\Delta E \Delta t = \hbar \qquad 53$$

ΔE is the line width and Δt the lifetime of the state, and so, for a typical γ-transition having an energy of 10 keV and a lifetime of 10^{-7} s the fractional line width is

$$\frac{\Delta E}{E} = \frac{4.6 \times 10^{-16}}{10^{-7} \times 10^4} = 4.6 \times 10^{-13} \qquad 54$$

Normally, however, the thermal motion of the atoms emitting the γ-rays broadens the line so that the fractional width actually observed is more like 10^{-6}. This is called thermal or Doppler broadening.

The second point of difference is in the actual energy of the γ-ray. This is less than the transition energy because some is taken by the recoil of the emitting nucleus. The amount of energy so lost is given by

$$E_R = \frac{E_\gamma{}^2}{2mc^2} \qquad 55$$

where E_γ is the energy of the excited level emitting the γ-ray and m is the mass of the nucleus. If the γ-ray had not lost some of its energy in this way it would have been possible for it to be absorbed by a nucleus similar to the one from which it was emitted, provided that the absorbing nucleus itself could not recoil and so take up extra energy. Such a process raises the second nucleus up to the same energy level as the original one and is called resonant absorption (p 30). Moon was able to observe resonant absorption in 1951 by making use of the Doppler shift. He mounted the emitter on the end of a rotor so that the additional velocity of the rotating source was sufficient to compensate for its loss in energy by recoil. This procedure does not, of course, reduce the line width, but merely makes the emission and absorption energy more nearly equal.

In 1958, however, Mössbauer discovered that, in some cases, if the γ-emitting atoms were held firmly in a crystalline lattice, a very narrow line with energy equal to E_γ and with thermal broadening suppressed was produced, because the large mass of the crystal (m in equation 55) absorbs the momentum of the recoil while taking up a minimal amount of energy. The γ-ray energy must be fairly low or lattice vibrations will be excited which will allow absorption of some of the energy. Mössbauer also discovered that recoilless absorption could take place as well as recoilless emission if the absorbing atom was similarly clamped in a lattice, the line width again being very narrow. Because the γ-line is so narrow a very small change in the energy level of the absorbing nucleus will prevent resonant absorption from taking place. This is the key to the importance of Mössbauer spectroscopy because such small energy differences can be caused by chemical and physical changes in the absorber environment and can therefore be used as an extremely sensitive tool for examining such changes.

ii Experimental equipment

Essentially, all that is required of the experimental equipment is that it is capable of slightly changing the energy of the γ-ray produced in a recoilless emission. This slight energy change is produced either by oscillating the absorber to and fro in front of the emitter or *vice versa*, so making use of the Doppler shift. In this way the extra velocity will either reduce or increase the γ-ray energy just sufficiently for it to match the energy of the level exactly and allow absorption to take place. A detector of some kind, which may be an ionization device, or a scintillation counter, measures the intensity of the γ-rays which pass through the absorber and this intensity shows a sharp reduction when resonant absorption occurs. *Figure 31* illustrates the lay-out of the equipment.

The oscillator, most frequently a high-fidelity loud speaker movement, is adjusted so that the velocity passes through that required for absorption and back again, the process being repeated continuously. The pulses from the detector are stored by a multi-scaling device, synchronized with the oscillator in such a way that those pulses leaving the source when it is moving at any particular velocity

FIG. 31.   Diagram of the essential parts of a Mössbauer equipment.

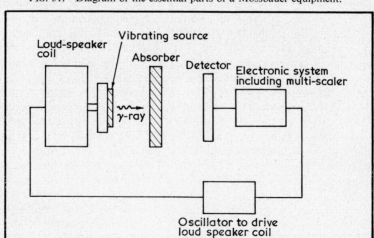

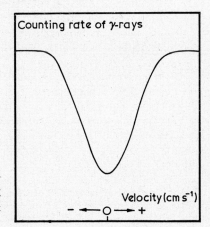

Counting rate of γ-rays

Velocity (cm s^{-1})

$-$ ⟵ O ⟶ $+$

FIG. 32. Simple Mössbauer spectrum. The number of γ-rays counted is sharply reduced as the velocity passes through that for resonant absorption.

are always stored in the same channel. The result is a spectrum such as that shown in *Fig. 32*.

There are many practical difficulties in making Mössbauer measurements. Examples of these are the fact that it is often necessary to cool absorber and source to liquid nitrogen temperatures so as to reduce the proportion of events which will excite vibrations in the crystal lattice, and the problems associated with producing uniform sources and absorbers. Such details will not be discussed here and the reader is referred to the bibliography at the end of the chapter for suitable texts.

iii *Mössbauer sources*

For a γ-ray transition to be suitable for Mössbauer work it must satisfy certain conditions of which the following are the most important.

(*a*) The γ-ray must be emitted from an excited state of a nucleus having a very short lifetime and this can be conveniently formed by α- or β-decay. This parent must have a half-life of at least a few hours to allow the measurements to be made. The excited state can also be formed by coulomb excitation.

(*b*) The transition must be to the ground state of the Mössbauer nucleus or it will not be possible to carry out the re-absorption of the γ-ray.

(*c*) The γ-transition energy must be reasonably low (say below 150 keV) so as not to excite lattice vibrations, but must not be too highly converted.

(*d*) The lifetime of the excited state must not be too short because, as the Uncertainty Principle shows, the shorter it is, the greater the energy spread.

3

In addition to these nuclear considerations, it is also essential that the source can be embedded in a lattice in a uniform way and that the resulting material is able to stand up to the low temperatures and to the movement of the oscillator.

At present some chemical aspects of about 15 elements have been examined by Mössbauer techniques and there are at least 20 more which have isotopes satisfying the nuclear requirements. The majority of the work has been carried out on tin and iron because these two elements have Mössbauer isotopes which have easily the best combination of characteristics. They have the properties shown in Table 5.

Table 5. Properties of iron and tin

Parent nucleus	Mössbauer nucleus	Transition energy (keV)	Half-life of parent
$^{57}Co \xrightarrow{\text{electron capture}} {}^{57}Fe*$	$^{57}Fe*$	14.4	270 d
$^{119m}Sn \xrightarrow{\text{isomeric transition}} {}^{119}Sn*$	$^{119}Sn*$	23.8	245 d

iv Applications of the Mössbauer method

From the viewpoint of the chemical applications of Mössbauer spectroscopy the important thing is the way in which the energy of the Mössbauer level in the nucleus is affected by chemical changes, i.e. by changes in the distribution of the electrons around the nucleus. One such effect is called the isomer (or chemical) shift.   A change in the s-electron density of the atom alters the proton distribution in the nucleus by means of a coulomb interaction.   The amount of this change is not quite the same for the excited and ground states of the nucleus and hence there is a slight change in the transition energy and the valley in the Mössbauer spectrum moves in one direction or the other.   The changes in electron density may be caused by a change in the valency state or in the bonding of the absorber atoms and the measurement can be used to give a quantitative estimation of such changes.

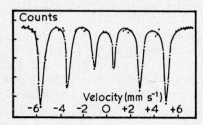

Fig. 33. Mössbauer spectrum of metallic iron showing hyperfine splitting of the ^{57}Fe spectrum. The six valleys correspond to the six transitions shown in Fig. 34. (Reprinted by permission, from Chemical and biological applications of Mössbauer spectroscopy, N. N. Greenwood, Endeavour, 1968, 27, 33.)

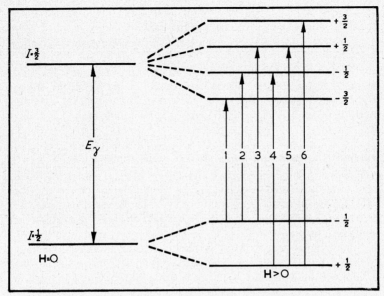

FIG. 34. Diagram showing how hyperfine splitting of nuclear energy levels in a field **H** > 0 gives rise to six possible transitions. Two others, having $I = 2$ are forbidden by selection rules. (Reprinted, by permission, from *Chemical and biological applications of Mössbauer spectroscopy*, N. N. Greenwood, *Endeavour*, 1968, **27**, 33.)

It is also possible for the Mössbauer level to be split into a number of sub-levels in which case the spectrum may look like the one shown in *Fig. 33*. This splitting can be caused by the coupling of the nuclear angular momentum with an internal or external magnetic field ('nuclear hyperfine splitting' or 'Zeeman splitting'). When this happens, a level with nuclear angular momentum I splits into $(2I + 1)$ sub-levels. If the ground state level has $I = \frac{1}{2}$, for example, this will split into $+\frac{1}{2}$ and $-\frac{1}{2}$, and if the Mössbauer level has $I = \frac{3}{2}$ this will split into $-\frac{3}{2}$, $-\frac{1}{2}$, $+\frac{1}{2}$ and $+\frac{3}{2}$. There are now eight possible transitions between the split Mössbauer level and the split ground state (*Fig. 34*). Two of these are forbidden by the selection rule which only permits changes $\Delta I = 0$ or 1, leaving six in all. This explains why the [57]Fe spectrum in *Fig. 33* shows six valleys. Zeeman splitting has been used to study magnetic fields in crystals and the properties of paramagnetic substances.

A second way in which the Mössbauer spectrum may be split is by a coupling of the nuclear quadrupole moment, which arises because the protons in the nucleus are not distributed symmetrically, with an electric field. Since the quadrupole moment is affected by the electron distribution of the atom the energy changes in the

Mössbauer spectrum can give information about this distribution as well as about the electric field in the lattice.

Additional information can be obtained from Mössbauer measurements by examining the effect on the spectrum of changes of temperature, by the relative intensities of the various lines, and by their widths. Taking all the possibilities into account, it can be stated that Mössbauer spectroscopy is one of the most important developments in structural chemical analysis that has taken place for a very long time.

Nuclear magnetic resonance spectroscopy (nmr)

Nuclear magnetic resonance spectroscopy, the principles of which were only discovered in 1946, has grown with great rapidity to the very important position it holds today. It is extensively used in structural chemical problems, in investigations into the course of chemical reactions, for analytical purposes and for examination of the magnetic properties of systems. It has the great advantage of being both rapid and non-destructive. This account of the subject will be restricted to the main principles behind the method and to a few examples of its use.

i *Principles of nmr*

As shown (p 2), any nucleus has a nuclear angular momentum represented by a quantum number I. In a uniform magnetic field this angular momentum can be orientated relative to the field in $(2I + 1)$ ways. These orientations are represented by the quantum number m_I which has values $I, I - 1 \ldots - I + 1, -I$. The magnetic moment associated with the nucleus is

$$\mu = g_I I \qquad\qquad 56$$

where g_I is the nuclear gyromagnetic ratio (p 2). The energy of this moment in a uniform field H_0 is

$$E = \mu H_0 = g_I I H_0 \qquad\qquad 57$$

The energy of a level with magnetic quantum number m_I in the field H_0 is therefore

$$E = g_I m_I H_0 \qquad\qquad 58$$

and so the energy difference between two levels having $\Delta m_I = 1$, is simply

$$\Delta E = g_I H_0 \qquad\qquad 59$$

Since ΔE is equal to $h\nu$, the frequency of this transition is

$$\nu = g_I H_0 / h \qquad\qquad 60$$

The angular momentum of the nucleus causes it to precess round the direction of H_0 with a frequency (the Larmor precession frequency), which can be found by classical methods to be

$$v = g_I H_0 / h$$ 61

This is the same as the transition frequency.

If an alternating field of frequency exactly equal to the Larmor frequency is applied at right angles to H_0 the nucleus absorbs energy equal to ΔE in equation 59. This causes its magnetic moment to change direction and m_I to alter by one unit. The process depends on the fact that the angular momentum can be orientated in more than one direction: for nuclei having $I = 0$ there is only one possible orientation, and such nuclei cannot therefore be used in nmr.

The reason why the method is so useful is that the field, H_0, at the nucleus on which ΔE, and hence the resonance frequency depends, is not quite equal to the actual applied field because of the screening effect of electrons present. Small changes in this electron arrangement change this field, and therefore the frequency, and hence the position of the resonance peak in the spectrum is altered. In addition to this,

FIG. 35. Diagram of an nmr spectrometer. The probe (in which the sample is placed) shown here is of the crossed coil type.

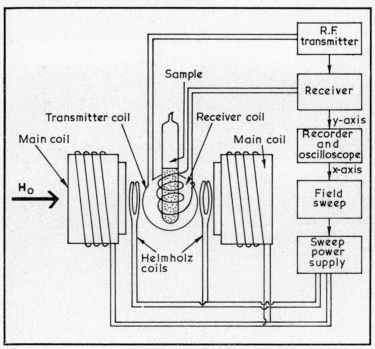

g_I the other parameter concerned in determining the frequency, has a value which is a property of a particular nucleus.

ii *Method of making the measurements*

The apparatus used is shown diagrammatically in *Fig. 35*, while *Fig. 36* depicts an actual high resolution machine. The homogeneous field, H_0, is provided by a very stable electromagnet and the sample is placed inside an rf oscillator coil between its poles. The oscillator frequency is kept constant, but the main magnetic field is swept through the resonance value, given by equation 60. The resonance point is detected by the sharp absorption of energy from the rf system which occurs when it is reached. The usual parameter measured is the chemical shift, *i.e.* the change in the resonance frequency due to a difference in the electronic structure of the materials being compared. Since the electronic structure is dependent on the actual chemical environment of the nuclei which are resonating, the position of the resonance peak is therefore sharply dependent on the chemical structure of the material. Although absolute measurements can be made, it is usual to compare the substance with a reference material, (usually tetramethylsilane in the case of proton measurements), and to quote the chemical

FIG. 36. Photograph of Varian HA-100D high resolution nmr spectrometer used at AERE, Harwell. (Reprinted, by permission of the UKAEA.)

shift, δ, in terms of the values of the field at resonance for the two samples. Thus

$$\delta = \frac{H - H_{ref}}{H_{ref}}$$

62

If the instrument is capable of high resolution the peaks may be seen to have a fine structure, *i.e.* to be split into several peaks. This is due to a coupling of the angular momenta of the two nuclei in a molecule through the medium of their bonding electrons and adds considerably to the structural information which can be obtained.

iii *Applications*

There are at least 100 nuclei which are suitable for nmr studies and the majority of elements possess at least one of them. However, for many elements the technique has only been used sporadically. Convenience is certainly one of the main reasons for this: it is, for

FIG. 37. Illustration of how a structural change in a compound is reflected in the nmr spectra. The upper trace is the spectrum of iso-butanol, the lower of n-butanol.

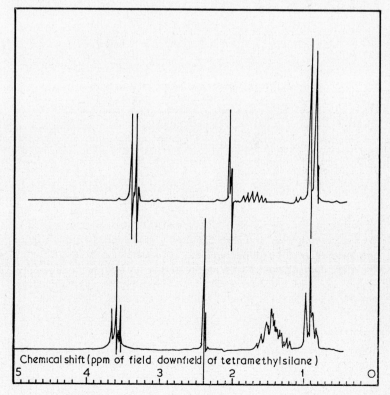

Chemical shift (ppm of field downfield of tetramethylsilane)

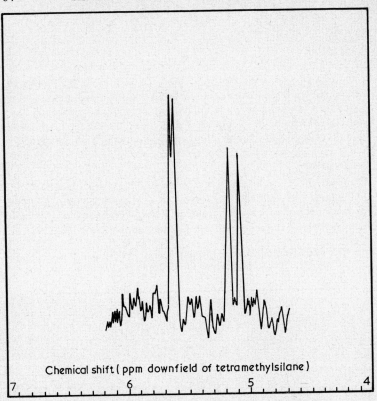

Chemical shift (ppm downfield of tetramethylsilane)

FIG. 38.   Another example, this time showing a more subtle difference between the compounds.   The trace is part of a spectrum of a mixture of isomers with two or more asymmetric carbon atoms.   The threo form has the larger chemical shift and gives rise to the doublet on the left, which has a splitting of 4 Hz.   The doublet on the right is due to the erythro form.

example, not very easy to use nmr for calcium compounds, since the only suitable isotope is ^{43}Ca with a natural abundance of only 0.13 per cent.   Hydrogen is one of the most useful nuclei because organic compounds can be studied, and this is made easier by the fact that ^{12}C and ^{16}O both have $I = 0$ and therefore do not give nmr spectra. *Figures 37* and *38* show examples of spectra which demonstrate how nmr can detect structural changes in material.

Positronium

Some of the material in this section will be concerned with the inorganic chemistry of positronium and should strictly be in Chapter 7, but it is included here so as to present a more readable account.

i *Formation and properties*

When a positron has been created (p 45) it is possible that, before it finally annihilates, it may form a system with a negatron in which the two particles rotate round their common centre of mass in a bound state for a comparatively long time—it may be as long as 1.4×10^{-7}s. Such an entity obviously resembles a hydrogen atom in some ways and is often referred to as the lightest known element, positronium with symbol Ps. This similarity even extends to its energy level system since it has an optical spectrum analogous to that of hydrogen, although no one yet has been able to observe any of its transitions.

Not all positrons form positronium before annihilation however, the alternative being slowing down and eventual 'free annihilation', that is, annihilation in a head-on collision with a negatron, sometimes after the intervention of other processes in which the positron becomes momentarily a part of the electron sheath of an atom. Because electrons have spin, two kinds of head-on collision can take place. When the spins are parallel it is a 'triplet' collision, when anti-parallel, 'singlet'. A singlet collision results in the production

FIG. 39. Diagram showing the relationship between the first ionization potential of the slowing down medium, I; the energy of its first excited level, E_x; and the threshold energy for positronium formation in argon gas. The values given are for positronium formation in argon gas.

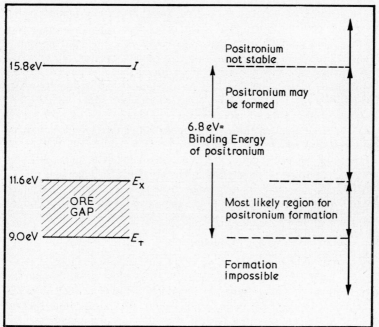

of two γ-rays each having an energy of 0.511 MeV. A triplet collision gives rise to three γ-rays having energies related to the angles between them but totalling 1.022 MeV. Singlet collisions occur 372 times as often as triplet collisions.

Positrons are slowed down in a medium by the same processes as negatrons (p 42) but when their energy has been reduced to that of the first ionization potential, I, of the molecules of the slowing down medium there is the possibility of abstracting a negatron from an atom and forming positronium. This happens to a sizeable fraction of the positrons, often as much as 30 per cent. This process remains possible until the positron energy drops below its threshold energy, E_T, which is equal to $(I - 6.8)$, 6.8 eV being the binding energy of the positronium atom: note that this energy is about half that of the hydrogen atom. Positronium formation is most likely when the energy of the positron is between E_T and E_X where E_X is the energy of the first excited level of the atoms of the medium. The reason for making this reservation is that above E_X many of the positrons will simply cause excitation of the atoms rather than form positronium. The region of most probable formation is called the Ore gap after the inventor of the model. *Figure 39* shows the relationship between the various energies. I and E_X depend on the slowing-down medium, for example, in the case of argon gas I is 15.8 eV and E_X is 11.6 eV. The Ore gap is therefore from 9.0 to 11.6 eV.

Depending on the orientation of the spins of the positron and negatron, positronium can be formed in either a singlet (p-positronium with opposed spins) or triplet (o-positronium with parallel spins) state. Formation of o-positronium is three times as probable as p-positronium. Furthermore, once formed it is much more stable, its annihilation only being 1/1120 as frequent. The lifetimes of these two forms in free space are 1.25×10^{-10} s for the p- and 1.4×10^{-7} s for the o-form. In practice, the lifetime of o-positronium may be shortened because of reactions with atoms or molecules in the medium, which cause quenching. This reduces the lifetime to some value between that of the two forms and may also alter the angular correlation of the annihilation γ-rays. Quenching is of two kinds, 'conversion' in which an o-positronium atom collides with a molecule and changes to the short-lived p-form, and 'pick-off' in which its positron annihilates with a negatron in a molecule instead of with its own.

ii *Experimental methods*

Three main types of measurement give information about the interactions of positronium in a medium. They are, the observed lifetime of the o-positronium, the amount of o-positronium formed under different conditions, and the angular correlation of the

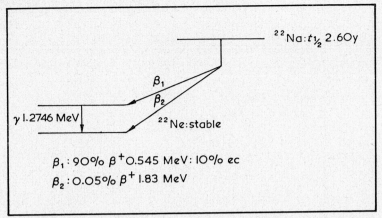

FIG. 40. Decay scheme of ^{22}Na : β^+ = positron emission, ec = electron capture.

annihilation γ-rays. A source of positrons is needed to make the measurements. ^{22}Na, half-life 2.6 y, has a decay scheme shown in *Fig. 40* and is so convenient that it has been used for the great majority of work. The 1.27 MeV γ-ray is emitted virtually simultaneously with the birth of the positron and is used as a time-marker for the start of its lifetime: the appearance of the annihilation radia-

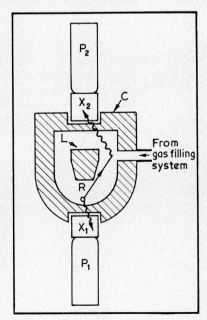

FIG. 41. Apparatus for observing positron lifetime. The source, R, emits a γ-ray which is detected by the scintillator X_1. L is a lead plug preventing the γ-rays reaching detector X_2 which can, however, detect the γ-rays from the positron when it finally annihilates. The source is inside a pressure vessel, C, which is filled with the gas under investigation. P_1 and P_2 are photomultipliers. (Reprinted, by permission, from 'Positronium', F. F. Heymann, *Endeavour* 1961, **20**, 225.)

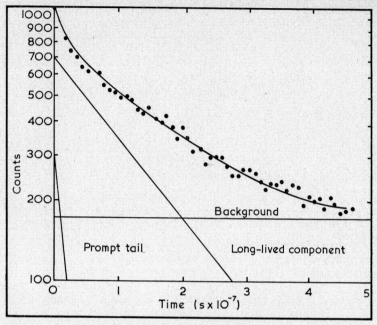

Fig. 42. Time distribution of positrons obtained from the apparatus shown in *Fig. 41.* The positron lifetime under the conditions of the particular measurement is obtained from the exponential curve marked 'long-lived component'. (Reprinted by permission from 'Positronium', F. F. Heymann, *Endeavour*, 1961, **20**, 225.)

tion indicates its death. A typical experimental arrangement for measuring this is shown in *Fig. 41.* The 1.27 MeV γ-ray recorded in scintillation detector X_1 starts a multi-scaling device which stops when the annihilation γ-ray appears in detector X_2, one event then being recorded at that time period. The process is repeated many times and so eventually a time distribution of the positrons is built up as shown in *Fig. 42.* The effect on this distribution of changing the medium in which the positrons annihilate can thus be studied.

The variation in the amount of *o*-positronium can be observed directly by surrounding the source with three counters at 120° to each other and recording the number of events in which a γ-ray is detected simultaneously in all three, while the angular correlation of annihilation γ-rays is measured by recording the simultaneous appearance of 0.511 MeV γ-rays in two counters and seeing how the number of such events varies as one counter is moved relative to the other.

iii *Chemical effects*

In essence, an experiment consists of an investigation of the amount of quenching of o-positronium caused in a particular medium under particular conditions. Because quenching makes the positronium undergo singlet (2γ) annihilation it reduces the lifetime from its maximum possible triplet (3γ) value and may also alter the angular correlation of the 0.511 MeV γ-rays, because this depends on the energy of the p-positronium when it annihilates. Experiments making use of this have examined the presence of free radicals in a reaction, since these may have a strong effect on quenching. Defects in crystal lattices caused by radiation also affect quenching, and changes in lifetime have also been observed coincident with phase transitions in crystals. Studies such as these are only in their early stages, but it seems probable that the method will develop into a powerful technique in these fields.

Apart from the use of positronium as a tool for such structural and reaction studies, its own chemistry has great fundamental interest. It behaves like hydrogen in many ways and formation of such structures as PsO_2, PsCl, PsI have been reported. There seems little doubt that this is going to develop into a wide area of work in the future.

Mesonic and other exotic atoms

Positronium, discussed in the previous section, is one of a number of bodies which have recently come to be called exotic atoms. Many of these atoms incorporate mesons of one kind or another in their structure and this section will consider two main types. In one (muonium and mesonium), a positive muon becomes the atomic nucleus, while in the other (mesonic atoms), a negative meson replaces an electron in the sheath of a normal atom which then becomes a muonic, pionic, or kaonic atom, depending on which kind of meson is present.

i Muonium and mesonium

Muonium, symbol Mu, is analogous to positronium in that it consists of a positive μ-meson and an electron. Like the latter, it can be regarded as being a light isotope of hydrogen and in its chemical and physical properties it can be compared with that element. It has a lifetime of 2.2×10^{-6} s.

Muons are formed by the decay of pions

$$\pi^+ \rightarrow \mu^+ + \nu_\mu \qquad\qquad 63$$

and have been found to be highly polarized, *i.e.* their spins are orientated so that the spin vector is in the opposite direction to the linear momentum of the muon when they decay.

$$\mu^+ \rightarrow e^+ + \nu_e + \bar{\nu}_\mu \qquad\qquad 64$$

($\bar{\nu}_\mu$, ν_μ and ν_e are various kinds of neutrino).   The positrons are preferentially emitted in the direction of the spin vector and so can be used as an indicator of the amount of polarization.   When muons interact with matter muonium may be formed, and if so half of it will be in a singlet and half in a triplet state (cf. positronium, p 65).   The singlet state is unpolarized and so there is a net reduction over the original polarization of the muons.

As in the case of positronium, quenching, that is the changing of triplet muonium into singlet, can occur with a corresponding reduction in polarization.   The amount of this quenching which takes place is dependent on the kind of reactions in which the muonium takes part, and hence a study of it gives information about these reactions.   Muonium is a hydrogen-like atom and so it should be chemically reactive and, in fact, studies of quenching by different materials have shown that it does form hydrides and takes part in many chemical reactions.   This work is still in its infancy and considerable developments may be expected when the various accelerating machines specifically designed to produce intense meson beams (the 'meson factories') have come into full operation.

The other exotic meson atom has been called mesonium.   It consists of a positive and a negative meson rotating round each other and may also be thought of as being a light hydrogen isotope. It has been theoretically predicted that such a system could form a bound state, but the practical difficulties inherent in producing a sufficient concentration of the two types of mesons in the same place at the same time has meant that it has yet to be seen experimentally.

ii *Mesonic atoms*

While other types of mesons can replace an electron in an atom, by far the most information is available about muonic atoms, the muon lifetime of 2.2×10^{-6} s making them easier to study.   The different types of mesonic atoms are very similar in general behaviour however, and so this section will be confined to a brief description of those which incorporate a muon.

Negative muons, formed from the decay of negative pions, may have a kinetic energy of hundreds of MeV and they lose most of it in the same way as do heavy charged particles (p 42).   Once their kinetic energy is gone they may be captured by an atom of the slowing-down material, which maintains its neutrality by losing an electron.   At first the muon rapidly loses energy in a 'cascade' down to lower states with the emission first of Auger electrons (p 29) and finally of radiation, the 'muonic x-rays'.   The muon mass is about 207 times greater than that of the electron and so when it finally attains the K-shell (the $1s$ orbital), the muon finds itself in an orbit with a radius 207 times smaller than that of the electron K-shell.

Eventually the muon will disappear, either by normal decay or by being absorbed by the nucleus.

On the chemical side, the energies of the muonic x-rays are, of course, characteristic of the chemical element concerned, and therefore measurements of their energies and intensities show how muons are being captured by the different elements in the compound. Changes in the chemical composition of the compound will be reflected in changes in the x-ray spectra and it seems probable that this type of measurement will become a useful structural chemistry tool when high intensity meson sources become more generally available.

Angular correlation of cascade γ-rays

If two γ-rays are emitted from a nucleus in a cascade, as shown in *Fig. 19*, the angle of emission of the second one will be correlated with that of the first, their relationship depending on the characteristics of the nuclear levels from which they have come. The nuclei in the source are orientated at random and so the γ-rays come from it isotropically, but if a particular angle is chosen for the first γ-ray only those nuclei orientated in a particular direction are selected. Under these conditions the angular correlation of the two γ-rays

FIG. 43. Angular correlation of ^{111}Cd cascade γ-rays showing the effect of changing the chemical state of the γ-emitting nucleus. (Reprinted by permission from 'Chemical and structural effects on nuclear radiations', S. de Benedetti, F. de S. Barros, and G. R. Hoy, *A. Rev. nucl. Sci.*, 1966, **16**, 31.)

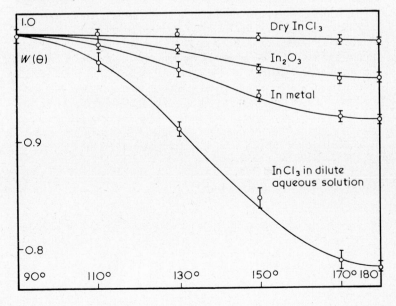

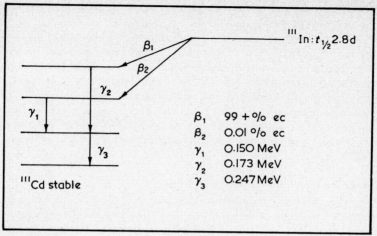

FIG. 44.   Decay scheme of ^{111}In.   Measurements shown in *Fig. 43* were carried out with the cascade γ-rays marked γ_2 and γ_3.

can be measured in a γ–γ coincidence experiment. The results of such measurements are usually given as a plot of the relative probability of the two γ-rays being emitted at a certain angle to each other against the angle.

The angular correlation depends on the angular momenta of the nuclear states involved in the γ-transitions and these momenta can be affected by electric and magnetic fields acting on the nucleus. This is why a change in the conditions in which the nucleus finds itself may be reflected by a change in the angular correlation.

Figure 43 shows very clearly that the chemical state of the compound containing the γ-emitting nucleus has a profound effect on the angular correlation of the cascade γ-rays. In this case the radioactive nucleus is ^{111}In whose decay scheme is shown in *Fig. 44*: the two γ-rays are marked γ_2 and γ_3. The method has not yet been extensively used in chemical investigations but the example given here indicates that it has considerable potential for structural chemistry and there seems little doubt that applications of the method by chemists will increase in the future.

Suggestions for further reading

X-ray diffraction

H. Lipson, 'X-ray crystallography', *Contemp. Phys.*, 1960, **1**, 370.

L. Bragg, 'X-ray crystallography', *Scient. Am.*, 1968, Jul. 58.

'Diffraction methods' by C. K. Prout in *Physical methods in advanced inorganic chemistry* (H. A. O. Hill and P. Day eds). London: Interscience, 1968.

Neutron diffraction

G. E. Bacon, 'The application of neutron diffraction', *Endeavour*, 1966, **25**, 129.

W. M. Lomer, 'Neutron spectroscopy of solids', *Contemp. Phys.*, 1966, **7**, 278; 1966, **7**, 401.

'Diffraction methods' by C. K. Prout, in *Physical methods in advanced inorganic chemistry* (H. A. O. Hill and P. Day eds). London: Interscience, 1968.

P. J. Wheatley, *The chemical consequences of nuclear spin*, ch. 6. Amsterdam: North-Holland, 1971.

'X-ray spectroscopy' by C. Bonnelle in *Physical methods in advanced inorganic chemistry* (H. A. O. Hill and P. Day eds). London: Interscience, 1968.

R. Jenkins and J. L. de Vries, *Practical x-ray spectrometry*, 2nd edn. London: Macmillan, 1970.

Mössbauer

W. E. Burcham, 'Nuclear resonant scattering without recoil', *Sci. Prog. Oxf.*, 1960, **48**, 630.

S. de Benedetti, 'The Mössbauer effect', *Scient. Am.*, 1960, Apr., 73.

R. L. Mössbauer, 'Recoilless nuclear resonance absorption', *A. Rev. nucl. Sci.*, 1962, **12**, 123.

P. B. Moon and D. A. O'Connor, 'Some applications of the Mössbauer effect', *Endeavour*, 1964, **33**, 131.

J. W. Smith, 'The Mössbauer effect and its chemical applications', *Sci. Prog. Oxf.*, 1966, **54**, 103.

N. N. Greenwood, 'Chemical and biological applications of Mössbauer spectroscopy', *Endeavour*, 1968, **27**, 33.

'Mössbauer spectroscopy' by J. Danon in *Physical methods in advanced inorganic chemistry* (H. A. O. Hill and P. Day eds). London: Interscience, 1968.

N. N. Greenwood, 'Applications of Mössbauer spectroscopy to problems in solid state chemistry', *Angew. Chem.*, 1971, **10**, 716.

P. J. Wheatley, *The chemical consequence of nuclear spin*, ch. 7. Amsterdam: North-Holland, 1971.

nmr

R. A. Y. Jones, 'Nuclear magnetic resonance', *Sci. Prog. Oxf.*, 1963, **51**, 198.

J. E. Ingram, 'Nuclear magnetic resonance', *Contemp. Phys.*, 1965, **7**, 13; 1965, **7**, 103.

J. E. Burgess and M. C. R. Symons, 'The study of ion-solvent and ion–ion interactions by magnetic resonance techniques', *Q. Rev. chem. Soc.*, 1968, **22**/3, 276.

'Nuclear magnetic resonance' by D. R. Eaton in *Physical methods in advanced inorganic chemistry* (H. A. O. Hill and P. Day eds). London: Interscience, 1968.

P. J. Wheatley, *The chemical consequences of nuclear spin*, ch. 8. Amsterdam: North-Holland, 1971.

Positronium annihilation and positronium formation

F. F. Heymann, 'Positronium', *Endeavour*, 1961, **20**, 225.

J. H. Green and J. Lee, *Positronium chemistry*. New York: Academic, 1964.

J. H. Green, 'Positronium formation and reactions', *Endeavour*, 1966, **25**, 16.

V. I. Goldanskii, 'Physical chemistry of the positron and positronium', *Atom. Energy Rev.* Vienna: IAEA, 1968.

A. J. Stewart and L. O. Roellig (eds), *Positronium annihilation*. New York: Academic, 1967.

H. J. Ache, 'The application of positron annihilation as a chemical probe', *Proc. 6th Int. Symp. on microtechniques*: Graz, 1970; Vienna: Verlag der Wiener, Medizinischen Akademie Wien, 1970.

Mesonic and other exotic atoms

V. W. Hughes, 'The muonic atom', *Scient. Am.*, 1966, Apr., 93.

C. S. Wu and L. Wilets, 'Muonic atoms and nuclear structure', *A. Rev. nucl. Sci.*, 1969, **19**, 527.

E. H. S. Burhop, 'Exotic atoms', *Contemp. Phys.*, 1970, **11**, 335.

G. Backenstross, 'Pionic atoms', *A. Rev. nucl. Sci.*, 1970, **20**, 467.

Angular correlation of cascade radiations

S. de Benedetti, F. de S. Barros and G. R. Hoy, 'Chemical and structural effects on nuclear radiations', *A. Rev. nucl. Sci.*, 1966, **16**, 31.

'Recent applications of perturbed angular correlations', *Hyperfine interactions* (A. J. Freeman and R. B. Frankel eds), ch. 13. New York: Academic, 1967.

S. G. Cohen, 'Hyperfine interaction and the angular distribution and correlations of nuclear γ-rays', ch. 12 in the above.

6. Analytical Chemical Applications

The counting of radioactivity

Many of the applications of nuclear effects to chemical problems require that the amount or energy of some kind of radioactivity be estimated. This section gives a summary of the methods of doing this, but does not attempt any detailed description of the counters or the way in which they operate since many good texts, some of which are listed in the bibliography, already exist on the subject.

The basic principle on which all counters depend is that the radiation interacts with the medium of the detector in the ways described in chapter 4 (p 42) and so produces ionization in it. In gas ionization counters and in semiconductor detectors the resulting charge is collected electrically, while in scintillation counters the ionization causes photon emission which is converted to electrical pulses by means of a photomultiplier. The primary ionization can also be detected by the trail it leaves in the medium as in cloud and bubble chambers, photographic emulsion, and solid track detectors, but counters based on these effects are of far less importance for most purposes than the others.

The kinds of counters most commonly used by radiochemists for α, β, γ, and neutron radiations and for fission fragments, are summarized in Table 6. Note that when energy resolutions are quoted these are for typical average conditions. (Resolution is defined as the full width at half maximum peak height of an energy peak divided by its energy and expressed as a percentage.)

Activation analysis

Activation analysis as a qualitative or quantitative procedure has grown steadily in importance with the increase in availability of sources of neutrons, γ-rays, and charged particles. It can be extremely sensitive: for example, under typical conditions, less than one ng of such important metals as As, Co, Cu, rare earths, Au, Mn, Hg, Ag and many others can be determined very rapidly, frequently without recourse to chemical separation. It is widely used in trace analysis, and is finding increasing application in the authentication of works of art and archaeological objects.

i *General principles*

The sample is bombarded with radiation which induces nuclear reactions in the nuclides present. Some of the reaction products

75

Table 6. Types of counters most commonly used for radiochemical measurements

Radiation	Type of counter	Used for energy measurement as well as simple counting	Remarks	Reference (see below)
α-particles	Gridded ion chamber	Yes: resolution 0.4%	Convenient	1, 2
	Proportional counter	Yes: resolution 1%	Rather awkward to use	1, 2
	Semiconductor detector (surface barrier)	Yes: resolution 0.2%	Very convenient	2, 3, 4
	Solid and liquid scintillators	No	Simple	2, 5
β-particles	Geiger-Müller counter	No	Simplest and cheapest	2
	Proportional counter	Yes: for low level energies, resolution 3%	More complicated than G.-M. counter for simple counting, but more accurate	1, 2
	Semiconductor detector (lithium drifted silicon)	Yes: resolution 1%	Very convenient for spectroscopy	2, 3, 4
	Solid and liquid scintillators	No	Simple	2, 5
γ- and x-rays	NaI (Tl) scintillator	Yes: resolution 6%	Cheap and convenient	2, 5
	Semiconductor detector (lithium drifted germanium or silicon)	Yes: resolution 0.3%	Rapidly becoming the most important counter for γ-spectrometry	2, 3, 4
	Proportional counter (Xe filled)	Yes: for x-rays: resolution 3%		6
	Solid and liquid scintillators	No	Simple	2, 5

Neutrons	BF$_3$ counter	No	For thermal neutrons	2
	Long counter	No	For neutrons 0.1–10 MeV: is a BF$_3$ counter surrounded by hydrogenous material	2
	^{6}Li spectrometer	Yes: resolution at 1 MeV 5%	For neutrons 0.02–1 MeV: complicated to use	7, 8, 9
Fission fragments	Semiconductor detector (surface barrier)	Yes: resolution 1%	Simple	2, 3, 4
	Solid track detectors	No	Very simple but time consuming	10

References: 1. W. Franzen and L. W. Cochran in *Nuclear instruments and their uses*, vol. 1 (A. H. Snell ed.), New York: Wiley, 1962. 2. J. Sharpe, *Nuclear radiation detectors*. London: Methuen, 1964. 3. G. Dearnaley and D. C. Northrop, *Semiconductor counters for nuclear radiations*. London: Spon, 1964. 4. G. Bertolini and A. Coche (eds), *Semiconductor detectors*. Amsterdam: North-Holland, 1968. 5. J. B. Birks, *The theory and practice of scintillation counting*. Oxford: Pergamon, 1964. 6. J. Du Mond, A. *Rev. nucl. Sci.*, 1958, **8**, 163. 7. M. G. Silk, AERE Report R5183, 1966. 8. M. G. Silk, AERE Report R5438, 1967. 9. M. G. Silk, AERE Report M2009, 1968. 10. R. L. Fleischer, P. B. Price and R. M. Walker, A. *Rev. nucl. Sci.*, 1965, **15**, 1.

are radioactive and their decay rate is proportional to the amount of product present (p 19), and hence to the amount of the original nuclide from which it was made. As shown earlier (p 21) the activity of a radioactive substance produced by irradiation in a flux F for a time t is

$$A = F\sigma N(1 - e^{-\lambda t}) \hspace{3cm} 65$$

where N is the number of atoms of the substance being irradiated, σ is the cross-section for the nuclear reaction, and λ is the disintegration constant of the radioactive product. In principle, N can be found by measuring A after irradiation for a known time t, but this assumes that σ is known and that F can be measured. This may not be the case, and there is also the additional difficulty that A is the disintegration rate of the sample and may be difficult to measure because the counter will not be 100 per cent efficient and because some of the activity may be lost by absorption in the sample itself. It is therefore usual to irradiate the sample and a standard of similar composition and containing a known amount of the required nuclide together, and count both of them under identical conditions. If this is done, all the above uncertainties are eliminated and the

Fig. 45. Cross-section for the reaction ^{197}Au$(n, \gamma)^{198}$Au. This is a typical neutron cross-section.

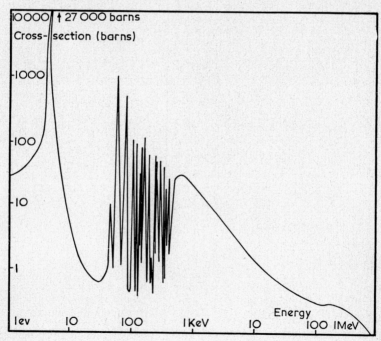

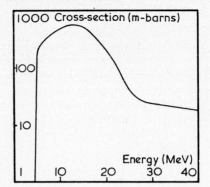

FIG. 46. Cross-section for the re-action $^{63}Cu(p, n)^{63}Zn$ showing the steeply rising threshold for this re-action.

proportions of the unknown and the standard are in the same ratio as their counting rates.

In deciding what nuclear reaction to employ it must also be remembered that the cross-section, σ, is not a constant but depends strongly on the energy of the bombarding particle. *Figure 45* shows a typical neutron cross-section, for the reaction $^{197}Au(n, \gamma)^{198}Au$ while *Fig. 46* is for the charged particle reaction $^{63}Cu(p, n)^{63}Zn$. This latter is a good example of the usual shape of the 'excitation function' for a charged particle reaction.

ii *Types of activation analysis*

Neutrons are the most frequently used bombarding particle for activation analysis. There are three main reasons for this. (*i*) They are much more readily available than charged particles; (*ii*) cross-sections for thermal neutron reactions are usually very much higher than for others and so sensitivities are increased, and (*iii*) unlike charged particles they penetrate deeply into the sample without difficulty. The commonest neutron reaction is the (n, γ) but with high energy neutrons a whole variety of others such as (n, p), (n, 2n), (n, α) *etc.*, may also occur and it is always necessary to be certain that the product being counted is produced only by the required reaction and not also by some other route from another nuclide present in the sample.

γ-rays can penetrate the sample even more easily than neutrons and provide an alternative for some analyses which cannot be performed with the latter, but the scarcity of suitable intense γ-ray sources militates against their wider use.

Charged particles suffer from the same disadvantage but they have been extensively used for analysis of light elements such as C, N and F which are not readily accessible to neutron activation techniques. They are readily stopped by the sample material (p 42) and this may result in problems of heat dissipation.

Finally, it should be mentioned that the type of prompt counting techniques used in so many 'on-line' physics experiments with accelerating machines may be employed in activation analysis. Nuclear levels excited in the bombardment of the sample may decay with very short half-lives ranging down to 10^{-14} s and the radiations emitted in their de-excitation may be counted while the sample is being irradiated. The advantage is that many new nuclear reactions become available and hence the scope of the activation technique is considerably widened. Many light elements have been determined in this way.

iii Sources of radiation
Table 7 gives a summary of the main types of radiation source used in activation analysis.

iv Counting the radioactive source
Any of the counters listed earlier in this chapter (p 75) can be used, but since most analyses are carried out on a comparative basis it is usually possible to count γ-rays. Lithium drifted germanium semiconductor detectors are normally employed for this and radiochemical purification of the sample can usually be avoided, although it will nearly always be essential if α- or β-rays are to be counted.

v Conclusions
Activation analysis is one of the most sensitive analytical procedures in use today and should always be considered when minor constituents are to be determined, but it must be remembered that there are many other methods for trace analysis. A comparison with some of these is shown in Table 8. Although published some time ago, the data in this table are still valid today and the convenience of some of the methods over activation may well make their use preferable.

Isotope dilution analysis
It sometimes happens that a quantitative estimation is required of a component of a complex organic or other mixture where it is difficult to purify the component without loss. In such a case it may be possible to perform the analysis by the isotope dilution technique.

A quantity of the same component is prepared pure, labelled with a radioactive isotope, and its specific activity (that is, its activity per unit weight) measured. A weighed amount of this is now added to the unknown mixture and steps are taken to ensure complete exchange of the activity between the added component and the unknown. The combined component is then purified, no account being taken of the yield, and its specific activity is determined.

Table 7. Radiation sources for activation analysis

Particle	Source	How the particles are produced	Particle energy	Typical flux	Remarks
Neutrons	Nuclear reactor	Nuclear fission	Thermal	10^{12} n cm^{-2} s^{-1}	The most convenient particle source
			Fast: average energy 0.1–0.5 MeV	10^{12} n cm^{-2} s^{-1}	
	Low energy accelerator, *e.g.* Cockroft-Walton	Nuclear reaction $^3H + {}^2H \rightarrow {}^4He + n$	~14 MeV	10^{10} n cm^{-2} s^{-1}	Deuteron energy required: 500 keV ∴ cheap small accelerator used
	Accelerator, such as van de Graaff or Cyclotron	Nuclear reaction $^9Be + {}^2H \rightarrow {}^{10}B + n$	Spectrum up to ~10 MeV	10^{10} n cm^{-2} s^{-1}	Deuteron energy required: 1 MeV ∴ expensive accelerator required
	Isotope mixture such as $^{241}Am + Be$	Nuclear reaction $^4He + {}^9Be \rightarrow {}^{12}C + n$	Average 4 MeV	2×10^6 n cm^{-2} s^{-1}	Convenient, but low flux
	Spontaneous fission source e.g. ^{252}Cf	Nuclear fission	Spectrum: average 1 MeV	10^9 n cm^{-2} s^{-1} per mg ^{252}Cf	Very convenient but not easy to obtain
Charged particles e.g. ^1H, ^3He, ^4He, etc.	Accelerator such as van de Graaff or Cyclotron	By electrostatic acceleration	Monoenergetic: energy set by accelerator conditions	10^{11}–10^{13} particles s^{-1}	High capital outlay
γ-rays	Accelerator such as electron linac	Bombardment of element by electrons	Broad spectrum	10^6 Röntgen min^{-1}	High capital outlay

Table 8. A comparison of some techniques of use in analysis of semiconductors*

	Several elements simultaneously	Sensitivity	Specificity	Accuracy	Freedom from contamination, reagent blanks, etc.	Possibility of overcoming surface contamination
Emission spectroscopy (inc. flame and atomic absorption)						
Direct	Yes	0.5–1 ppm	Good	Needs authentic standards of similar material	Good	Good
After chemical concentration	Yes	0.01 ppm	Good	Reasonable	Bad	Bad
Mass spectroscopy						
Vacuum spark	Yes	0.01 ppm	Good	Needs authentic standards	Good	Good
Isotope dilution	No	Very high	Good	Reasonable	Bad	Bad
Polarography	Few	0.1 ppm	Reasonable	Good	Bad	Bad
Vacuum fusion	O, N, H	1 ppm	Reasonable	Reasonable	Good	Bad
Colorimetric, fluorimetric	No	0.1 ppm	Reasonable	Good	Bad	Bad
Radiotracers	No	High	Good	?	Good	Bad
Activation	In some cases	Very high	Good	Good	Good	Good

* From A. A. Smales, *J. elect. Soc.,* 1955, **1**, 327.

If the amount of the component to be determined is M, the amount added is m, and the specific activities are A and a respectively, then the specific activity of the mixture, A, is $am/(M + m)$. From this it can easily be seen that the required amount, M, is equal to $m(a/A - 1)$.

The method allows determination of substances at levels of less than one ppm in some cases.

Dating by nuclear methods

The three main areas in which dating by nuclear methods has been applied are geology, archaeology and art. In geology the date of some drastic occurrence such as the solidification of the piece of rock in question can be found, in archaeology it is the date at which an artifact was made which is revealed, while in art it may be possible to date a painting or other object and so expose a forgery.

There are at least 20 different nuclear reactions which can be used in dating, but the general principle behind them all is that a radioactive substance decays at a constant rate (p 19) and so constitutes a kind of clock. Three of the more important ones will now be described in order to illustrate the method.

i Radiocarbon dating

This is by far the most important nuclear way of dating archaeological objects. The half-life of ^{14}C, on which it depends, is 5730 years and this means that the method covers the period from a few hundred to about 60 000 years ago, an ideal span for historical studies. The way in which it works is as follows:

1. Neutrons produced in the atmosphere by the action of cosmic rays react with nitrogen by the $^{14}N(n, p)^{14}C$ process.

2. The ^{14}C exchanges with normal ^{12}C in CO_2 and, at least until this century, this resulted in the amount of ^{14}C in atmospheric CO_2 being constant, one gramme of carbon emitting 15 particles per minute from decay of the ^{14}C.

3. Living tissue, thus had this constant 'specific activity' until the animal or plant died, at which point it took in no more CO_2 and so the specific activity started to fall.

To carry out the dating procedure the carbon in the specimen is usually converted into either CO_2 or CH_4, which is then counted in a gas or scintillation counter, the main difficulties being the avoidance of contamination with modern carbon and the fact that the isotope emits a rather soft β-particle with an energy of only 0.160 MeV.

The commonest materials to have been dated in this way are charcoal and wood, but leather, cloth, bone, horn and many other substances have also given satisfactory results.

ii *Potassium–argon dating*

This is given as an example of a dating technique suitable for geological purposes. It is based on the radioactive decay of ^{40}K to ^{40}Ar by electron capture. At some time the rock containing the potassium solidified. From that moment the ^{40}Ar daughter cannot escape and so the ratio $^{40}Ar/^{40}K$ gradually increases from its initial value of zero and can be used to date the occurrence. The half-life of ^{40}K is 1.3×10^9 years and so this clock is suitable for virtually the whole geological time span.

iii *Fission track measurement*

This is one of the newest and most promising methods. It depends on the fact that fission fragments cause physical damage to the material in which the fission occurs and that in many substances the damage track is permanent and can be made visible to an optical microscope by etching, or viewed directly with an electron microscope.

^{238}U is frequently present in geological materials and also in artifacts like glass and ceramics and it is clear that the density of fission tracks in such materials is dependent on the length of time since the object was made, as well as on the spontaneous fission half-life of ^{238}U (7×10^{17} y). The procedure has been successfully applied to samples ranging from glasses with a high uranium content manufactured less than 100 years ago, to minerals at least 10^8 years old. *Figure 47* shows examples of fission tracks in plastic, glass and mineral orthoclase.

FIG. 47. Fission tracks in various materials: left to right, in a polycarbonate plastic film, in glass, and in orthoclase. (Reprinted, by permission, from the article by R. L. Fleischer in *Radiometric dating for geologists*, I. Hamilton, (ed.). New York: Interscience, 1969.)

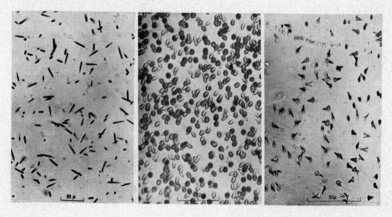

Forensic applications

While not a very large part of forensic science, there have been applications of neutron activation analysis and of radioactive tracers in crime detection. Undoubtedly there would be more of them if they were cheaper, but in spite of the cost, the employment of such methods is slowly increasing.

i *Neutron activation analysis*

Because of its high sensitivity for many elements, the activation technique can often reveal whether or not two specimens have a common origin. For example, a paint scraping may be compared with the paint on a particular motor-car by activation of the two samples to determine whether or not the trace elements are present in equal proportions in both. In the American courts, evidence of this kind has been admitted for paint, grease and other specimens from cars, for soil, and for illicitly distilled liquors, and several convictions have resulted. Work to detect common origin has also been conducted with drugs, glass, inks and other substances.

Hair is another type of material on which considerable research has been carried out. By examining a single hair one may be able to find out whether arsenic has been administered to the subject and even how often and at what intervals of time. Similarly, antimony and barium present on a hand which has recently fired a gun can be detected by covering the hand with paraffin wax, removing the wax and activating it. Again, evidence on these lines has been accepted in the American courts.

ii *Radioactive tracers*

It is sometimes difficult to photograph fingerprints by conventional methods on certain types of surfaces. However, if SO_2 with ^{35}S incorporated in it is applied to the surface, it reacts with the sweat residues of the fingerprint and the ^{35}S is deposited on the pattern. An 'auto-radiograph' may then be made of the print by simply placing a piece of film on it, the radioactivity causing fogging of the film.

Natural radioactivities have also begun to play a part, this time in the detection of art forgeries. Paintings contain lead in the form of white lead pigment. The ores from which the lead is extracted contain uranium, one of whose daughters is ^{226}Ra which itself eventually decays to ^{210}Pb which, in old ores, will be in secular equilibrium (p 21). On extraction the ^{210}Pb remains with the lead while much of the ^{226}Ra is removed. The ^{210}Pb now starts to decay with its 22 year half-life and the amount of it relative to the amount of ^{226}Ra present reveals whether the lead was processed recently or not and hence whether the pigment is new or old.

Application of nmr to analytical problems

The principles of nmr and its use in structural chemistry have been described earlier (p 60); a short paragraph is included here to show how the techniques may be applied in analytical chemistry.

The position of a resonance peak in an nmr spectrum is dependent on the particular structure in which the resonating atoms are bound and so the relative chemical shifts, expressed in accordance with equation 62, are constant for particular groupings of atoms and may therefore be used qualitatively to show the presence of a group; tables of standard shifts exist for many nuclides for this purpose. Furthermore, the area under the resonance peak is directly proportional to the number of resonating nuclei and so a quantitative estimate of this number can be made. Once again, while it is theoretically possible to make this measurement absolute, it is far more convenient and accurate to perform it relative to a standard.

X-ray spectroscopy for qualitative and quantitative analysis

When using x-ray spectroscopy in structural analysis one is concerned with the slight changes in the energy of characteristic x-ray lines caused by a change in the electron distribution of the emitting substance (p 53). For qualitative analytical purposes one makes use of the fact that the frequency (or energy) of a line is a fundamental property of a particular element, in accordance with Moseley's law (p 53), while the measurements can be used quantitatively because the intensity of the line is proportional to the number of atoms which are emitting the particular x-rays concerned.

The x-ray fluorescence technique is nearly always used; that is, the sample is bombarded with high intensity x-rays which excite its atoms and cause them to emit their characteristic x-rays. An alternative is to excite these x-rays by bombarding the sample with electrons, in which case the procedure is called electron microprobe analysis. The actual spectroscopy of the x-rays is carried out by means of a crystal spectrometer or by a semiconductor counter as described earlier (p 53).

X-ray fluorescence analysis is comparable with arc-spark emission spectroscopy in its general characteristics, but it does have the advantage of being non-destructive. Both techniques can determine a large number of elements simultaneously whether as major or minor constituents, but the resulting x-ray spectra are much the simpler of the two; it is, however, difficult to analyse for elements below fluorine in the Periodic Table by x-ray fluorescence. Comparison with standards is the normal procedure in both cases. Both require comparatively short times for the analysis which can be made automatic if required, and their costs are comparable.

Electron microprobe analysis differs from x-ray fluorescence in that the electron beam can be focused down to an exceedingly small, high intensity spot. This makes the method very useful for examining irregularities in composition of a surface and the micro-probe is unique in its ability to make measurements on volumes as small as μm^3. The corollary to this is, however, that it is less accurate than x-ray fluorescence in making average determinations of elements in a whole sample.

Suggestions for further reading

Counting of radioactivity

'Methods of detection and measurement of radioactive nuclides' by J. G. Cuninghame in *Radiochemical methods in analysis* (D. I. Coomber ed.). London: Plenum, 1973.

Activation analysis

W. S. Lyon (ed.), *Guide to activation analysis*. Princeton: van Nostrand, 1964.

R. F. Coleman and T. B. Pierce, 'Activation analysis', *Analyst*, 1967, **92**, 1.

W. H. Wahl and H. H. Kramer, 'Neutron activation analysis', *Scient. Am.*, 1967, Apr., 68.

M. Rakovic, *Activation analysis*. London: Iliffe, 1970.

Isotope dilution analysis

J. Ruzicka and J. Stary, 'Isotope dilution analysis', *Talanta*, 1964, **11**, 691.

T. T. Gorsuch, *Radioactive isotope dilution analysis*, Radiochemical Centre Amersham Pamphlet, 1967.

Dating by nuclear methods

E. I. Hamilton and R. M. Farquhar (eds), *Radiometric dating for geologists*. London: Interscience, 1968.

Forensic applications

D. E. Bryan and V. P. Guinn, *Use of neutron activation analysis in scientific crime detection*. J. J. Hopkins Laboratory Report, GA 6152, 1965.

W. S. Lyon and F. J. Miller, 'Forensic applications of neutron activation analysis', *Isotopes Radiat. Technol.*, 1966–7, **4**, 325.

R. F. Coleman, 'Nuclear techniques in forensic science', *J. Br. nucl. energy Soc.*, 1967, **6**, 134.

F. J. Miller, E. V. Sayre and B. Kersch, *Isotopic methods of examination and authentication in art and archaeology.* Oak Ridge National Laboratory Report, ORNL-IIC-21, 1970.

Applications of nmr to analytical problems

'Analytical application of nuclear magnetic resonance' by H. S. Gutowsky in *Physical methods in chemical analysis*, vol. III (W. G. Berl, ed.). New York: Academic, 1956.

Application of x-ray spectroscopy and diffraction

'Fluorescent x-ray spectrometric analysis' by G. L. Clark in *Physical methods in chemical analysis*, vol. III (W. G. Berl, ed.). New York: Academic, 1956.

J. G. Brown, *X-rays and their applications.* London: Iliffe, 1966.

7. Inorganic Chemistry

Chemical effects of nuclear transformations

As shown (p 16), when a nuclear transformation, such as emission of a particle, takes place energy is released. If, for example, an α-particle is emitted it will have a kinetic energy of several MeV and may itself cause chemical effects in the medium through which it passes. The study of the chemical effects caused by ionizing and other radiations is called radiation chemistry and is a large subject in its own right, but as it is not really an aspect of the atomic nucleus it will not be discussed here.

Not all of the energy of the nuclear transformation instanced above is taken away as kinetic energy of the α-particle however. The nucleus itself recoils with an energy

$$E_R = E\alpha \; \frac{m_\alpha}{m_D} \qquad\qquad 66$$

where m_α and m_D are the masses of the α-particle and of the daughter nucleus. It is this residual nucleus which will be considered here, because although by nuclear standards the energy given it by most kinds of nuclear transformation is not especially high, in chemical terms it is enormous: >1 keV compared with chemical bond energies of the order of 5 eV.

This high energy may cause chemical changes to take place in the atom or molecule of which the nucleus is part: for example, very highly reactive free carbon atoms are formed in reactions such as $^{12}C(n, 2n)^{11}C$ and enable the atomic reactions of this element to be studied. The chemistry of these highly excited recoil products is usually called hot atom chemistry. Its genesis is to be found in the experiments carried out by Szilard and Chalmers in 1934 when they discovered that the bond between carbon and iodine was broken if ethyl iodide was irradiated with thermal neutrons. Before discussing hot atom chemistry further, however, some of the ways in which the energy can be given to atoms by nuclear transformations will be considered.

i *The nuclear processes*

The recoil energy may come either as a result of a nuclear reaction or from radioactive decay of the nucleus. In the original Szilard–Chalmers process the nuclear reaction concerned was radiative neutron capture.

$$^{127}I(n, \gamma)\, ^{128}I$$

89

If a γ-ray of energy E_γ is emitted in such a process the recoiling nucleus is given energy equal to

$$E_R = 536E_\gamma^2/m_D \qquad\qquad 67$$

for each photon, although the final recoil energy may be a resultant of several such processes. Note that the atomic number of the atoms involved in the transformation does not change in the (n, γ) reaction, and this is also the case with the (γ, n), (n, 2n), (d, p), (p, pn) reactions and with radioactive decay by internal conversion. This means, of course, that the 'cold' chemistry of the reactant and product is the same. In the case of such reactions as (n, p), (n, α), and decay by α- or β-emission or by electron capture, however, the atomic number changes and this may be an advantage. For example, if the product is formed in a mixture of several possible valency states, there is more chance that these will remain unchanged until they can be examined if there is no large quantity of the same element present.

ii *Hot atom chemistry*

In thermal terms the atoms concerned are certainly hot. Nuclear reactions are often studied over an energy range from their threshold up to very high excitations, but it is not easy to put enough energy into the atom by ordinary means to do this for chemical reactions, and so experiments are conducted near to the reaction thresholds which as mentioned above, are typically of the order of 5 eV. The reason why hot atom chemistry is interesting, therefore, is because it does indeed look at the energy range above this.

Of course, the initial energy of the recoil atom is usually far too high for chemical reactions to take place and it loses most of this, first by ionizing interactions with the medium through which it is passing and then through electronic exchange and scattering reactions.

Normal chemical means are used to identify the products of hot atom reactions; for example, solvent extraction may serve to distinguish between the valency states of halogens produced in the irradiation of organic halides. The fact that the product is frequently radioactive greatly assists in its quantitative and qualitative analysis.

The hot atom reactions studied have been legion. Starting with the original Szilard–Chalmers reaction, halide systems, both organic and inorganic, have received extensive attention. Reactions in a large number of inorganic crystals, *e.g.* permanganates, and in those containing complex ions have also been closely examined, while the recoil chemistry of tritium in organic compounds has now come to have great importance because it is reasonably simple and can be used as a model for other hot atom reactions. There seems little point in dealing more fully with particular experiments in this monograph, if only because they embrace such an enormous volume of work. Fortunately there have been a number of reviews and symposia on the subject and these are listed at the end of this chapter. The field is still extremely active and it is to be expected that it will continue to contribute to our understanding of chemical reactions, particularly perhaps to those taking place in the solid state.

Isotope effects

i Definition

In addition to the chemical changes which may be produced by a nuclear transformation, there is another nuclear property which has a direct influence on chemistry. This is simply the nuclear mass. The changes which occur in some physical and chemical properties of atoms or molecules when their isotopic composition is altered are called isotope effects. These effects depend on the percentage change in the mass and are therefore much more marked for the lighter elements, being greatest for the hydrogen isotopes.

ii Chemical isotope effects

The most important chemical isotope effects are those concerned with equilibrium constants and with reaction rates. Both are due to changes in the energies of molecular quantum states, particularly the vibrational energies.

The vibrational energy of a diatomic molecule depends on the factor $1/\sqrt{\mu}$ where μ is the reduced mass of the molecule. If the atoms have masses m_1 and m_2, μ is equal to $m_1 m_2/(m_1 + m_2)$. Clearly then, if there is an isotopic change in one of the atoms μ will have a new value and so, therefore, will the vibrational energy. This will have an effect on both the equilibrium constants and the rates of reactions in which the molecule takes part.

iii Separation of isotopes

Differences between some physical property of the isotopes of an element are often used to effect their separation. The most obvious method is by means of a type of mass spectrometer called an isotope separator. Complete separations of isotopes on a g scale are quite easily achieved by such machines.

Other types of separation are statistical in nature. They rely on small differences in some bulk property of the isotopes concerned, such as vapour pressure or viscosity, and usually have to be operated in a series of stages, a small degree of enrichment being achieved at each stage. Examples of such methods are thermal diffusion, fractional distillation and centrifuging.

How the elements are synthesized in stars

Any theory of how the elements are formed in the cosmos must clearly be able to explain their observed relative abundances. *Figure 48* shows abundance data which are the results of measurements of meteorite composition and of stellar spectra: it will be discussed in the next section (p 97) and at this stage it is only necessary to note the following important points.

(*a*) Hydrogen and helium are by far the most abundant elements.

(*b*) While there are many irregularities, the general trend is an exponential decrease of abundances to about mass 100, after which they decrease very slowly.

(*c*) There is a large peak at iron.

(*d*) Even-A nuclides are more abundant than odd-A ones.

(*e*) Deuterium, lithium, beryllium and boron have much lower abundances than neighbour elements.

(*f*) Although it cannot be seen from the figure, neutron-rich nuclides predominate among the heavier elements.

Of course, data such as these are essentially local, because the necessary measurements cannot readily be made on distant stars. No one knows whether abundances are constant throughout the cosmos and indeed, there seems to be definite evidence that the heavy element composition of stars is not always the same, but the important thing at this stage must be to devise a theory of the origin of the elements which will at least explain our local findings.

i *The theory of stellar nucleogenesis*

There have been many attempts to explain the abundances but the only one which has come near to success is the proposal that the elements are formed in a series of nuclear reactions taking place in the interior of evolving stars. While many people contributed to the early development of this theory, it was first put forward in a complete form by Burbidge, Burbidge, Fowler and Hoyle in 1957. The various processes described in their lengthy paper still form the basis of the scheme which is used today and which will now be outlined.

There are many different possible reactions which can occur as a star evolves, but by far the largest part of the elements are formed by the nine main ones shown diagrammatically in *Fig. 49* and outlined in the following stages:

1. The star is formed by the condensation of gas which may be pure hydrogen or a mixture of hydrogen and helium. Gravity causes it to contract and, as it does so, it heats up until the hydrogen can enter into nuclear reactions which lead to the formation of helium.

This may come about either directly

$$^1H + {}^1H \rightarrow {}^2H + \beta^+ + \nu \qquad 68$$

$$^2H + {}^1H \rightarrow {}^3He + \gamma \qquad 69$$

$$^3He + {}^3He \rightarrow {}^4He + {}^1H + {}^1H \qquad 70$$

or, when a little carbon is present from some other process, through the 'carbon–nitrogen cycle' in which carbon and nitrogen act as catalysts.

$$^{12}C + {}^1H \rightarrow {}^{13}N + \gamma \qquad 71$$

$$^{13}N \rightarrow {}^{13}C + \beta^+ + \nu \qquad 72$$

$$^{13}C + {}^1H \rightarrow {}^{14}N + \gamma \qquad 73$$

$$^{14}N + {}^1H \rightarrow {}^{15}O + \gamma \qquad 74$$

$$^{15}O \rightarrow {}^{15}N + \beta^+ + \nu \qquad 75$$

$$^{15}N + {}^1H \rightarrow {}^{12}C + {}^4He \qquad 76$$

These two processes of conversion of hydrogen to helium are referred to as hydrogen burning and occupy the star for most of its life.

2. When the hydrogen has all been burnt, gravitational collapse again raises the temperature of the star until 'helium burning' can begin. In this phase, which is much shorter in duration than hydrogen burning, ^{12}C and ^{16}O are formed.

$$^4He + {}^4He + {}^4He \rightarrow {}^{12}C \qquad 77$$

$$^{12}C + {}^4He \rightarrow {}^{16}O \qquad 78$$

3. With the helium burnt, the temperature rises again and events move forward even faster. Two processes, 'carbon burning' and the 'α-process' begin. ^{20}Ne and ^{24}Mg are formed, and α-particles which are ejected from ^{20}Ne by high energy photons take part in further nuclear reactions

$$^{12}C + {}^{12}C \rightarrow {}^{24}Mg + \gamma \qquad 79$$

$$^{16}O + {}^4He \rightarrow {}^{20}Ne \qquad 80$$

$$^{24}Mg + {}^4He \rightarrow {}^{28}Si \ etc. \qquad 81$$

4. Further contraction and still higher temperatures now give rise to the 'e-process' on an even shorter time scale. A thermodynamic equilibrium is set up between the nuclides, protons, and neutrons present, with the result that the most stable nuclides are formed in the highest yield, so giving rise to the iron peak at ^{56}Fe, the nuclide with the highest binding energy of all.

5. There is no conceivable mechanism by which the remainder

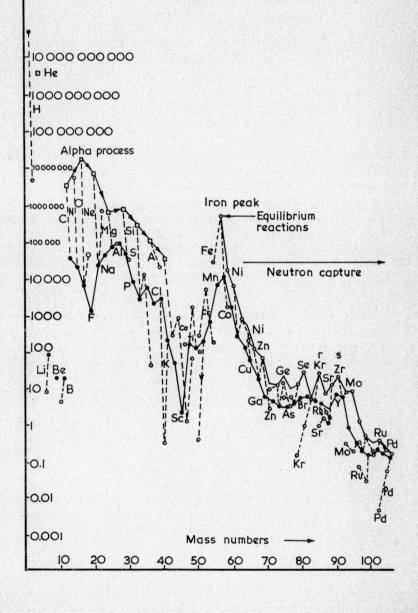

Fig. 48. The abundances of the nuclides. Black dots are for nuclides with odd-*A*, open circles, even-*A*. Dotted lines link isotopes of an element. Squares show elements which can be broken up into α-particles. Some of the nucleosynthetic processes are also indicated on the diagram. The data are those of Suess and Urey, compiled from meteoritic and spectroscopic data in 1956. (Reprinted, by permission, from 'Origin of the elements', B. Pagel, *New Scientist*, 1965, 8 Apr., 103.)

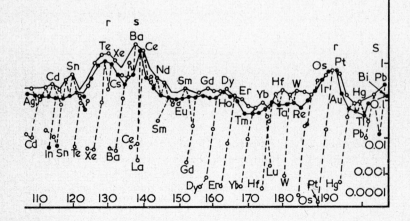

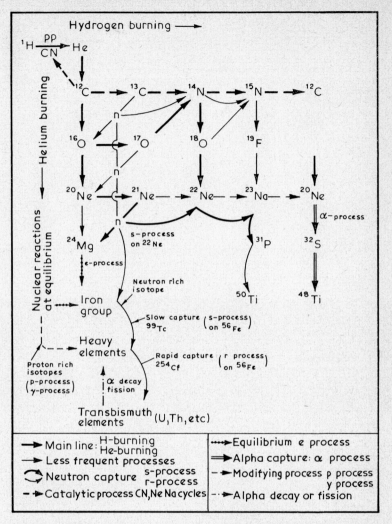

FIG. 49.   Synthesis of the elements in stars.   Elements synthesized by hydrogen burning reactions are listed horizontally, by helium burning and more complicated reactions vertically.   Curved lines show neutron capture reactions. (Reprinted, by permission, from 'Synthesis of the elements in stars', E. M. Burbidge, G. R. Burbidge, W. A. Fowler and F. Hoyle, *Rev. mod. Phys.*, 1957, **29**, 547.)

of the elements can be produced in this particular star which now has the option of contracting until it is a white dwarf or exploding as a supernova.   The supposition is that supernovae eject matter which is received by young stars or other celestial objects and that nuclear reactions take place in these.

In the 'r-process' the supernova matter is bombarded by neutrons arising from such reactions as

$$^{13}C + {}^{4}He \rightarrow {}^{16}O + {}^{1}n \qquad\qquad 82$$

$$^{17}O + {}^{4}He \rightarrow {}^{20}Ne + {}^{1}n \qquad\qquad 83$$

$$^{21}Ne + {}^{4}He \rightarrow {}^{24}Mg + {}^{1}n \qquad\qquad 84$$

at a neutron intensity so high that neutron capture reactions can take place much more rapidly than β-decay, and so bring about the formation of very heavy nuclides.

The 's-process' is similar but takes place on a slow time scale and this allows far more β-decay and so leads to the formation of elements from iron to polonium.

6. Finally, protons, perhaps produced in such reactions as

$$^{14}N + {}^{1}n \rightarrow {}^{14}C + {}^{1}H \qquad\qquad 85$$

take part in the 'p-process' to produce the few proton rich heavy nuclei, while the 'x-process', about which little is known, is invoked to explain the abundances of deuterium, lithium, beryllium and boron which cannot be accounted for by the scheme given above, because they are so easily destroyed by thermonuclear reactions at the high temperature prevailing in the evolving star.

ii Conclusions

The stellar nucleosynthesis programme outlined above does seem to be capable of reproducing the observed abundances very well, certainly better than any other so far. Of course, it is not the only way in which stars evolve, nor are ordinary stars necessarily the only bodies in the universe capable of synthesizing the elements, but it does at least account for a large part of the known facts. This is a field where new and interesting developments are constantly taking place and it is impossible to predict what will happen next. One can only watch with interest.

The abundance of the elements
i Sources of abundance data

There are three main sources of information on element abundances: stellar spectra, measured abundances of objects falling on the earth from space, and the earth itself.   To these one should add two others which are at present of lesser importance, examination of specimens brought back from the moon, and the analyses of the composition of cosmic rays.

Abundance tables normally give the so-called cosmic abundances, generally taken to mean abundances of the elements within our own galaxy. In fact, really detailed spectral information comes only from the sun, that from the stars in the remainder of the galaxy being decidedly more sparse, while spectral data from other galaxies are sketchy indeed. Analysis of meteorites provides the other major source of cosmic abundances. These objects consist of debris from the solar system, but a difficulty arises in the interpretation of the results because any particular meteorite may not be truly representative of the solar matter: its composition may have become altered by cosmic ray bombardment or by some physical process between the time of its birth and its arrival here. The class of meteorites called chondrites is generally thought to be the closest to the original matter and, as shown later, there is reasonable agreement between their composition and that of the sun.

There is, of course, an inexhaustible supply of material for the measurements from the earth itself. Unfortunately, it is not at all easy to use data obtained from this source in the preparation of the cosmic abundance table because of the extreme difficulty in sampling. How is one to relate the results of measurements on different types of rock, waters, crust and core to an average composition of the whole earth? A similar difficulty exists in the interpretation of analyses of lunar samples.

Cosmic ray abundances are in a different category from the others because, whilst their origin is still not really known, it is certain that a considerable proportion of them come from outside the solar system or even from outside the galaxy. Their composition is definitely different from the cosmic abundances but the significance of this fact is not properly understood at present.

ii Cosmic abundances

Some of the abundances shown in *Fig. 48* have been modified in the light of more recent measurements, but these changes do not alter the general picture given by the figure and do not affect the conclusions which were drawn about stellar nucleosynthesis. Table 9 is a list of some current values. In it the figures are given in 'cosmic abundance units', that is, they have been normalized to a value of 10^6 atoms for silicon: the choice of silicon as a datum level is a little unfortunate because this element is difficult to measure spectroscopically in the sun, and because there is some evidence that it may vary rather more than other non-volatile elements in cosmic debris. The first four columns of the table show mean values of measured abundances from the sun, from two types of chondrites and from the moon, while the last three are evaluated by various authors from all types of measurements.

The stability of matter and the double-humped potential barrier

It would, perhaps, have been more logical to include this section in Part 1, since it is really basic nuclear physics but it was decided to introduce it here because it is so interwoven with the subject matter of the remainder of this chapter that it seems essential to have it readily to hand.

i *Normal decay modes and the N–Z chart*

It was shown earlier (p 14) that there is only a narrow range of combinations of neutrons and protons in a nucleus within which stable nuclides are found, and it has been explained how this stability is a result of the interaction of the various forces within the nucleus. If enough extra neutrons or protons are added to a stable nucleus it becomes unstable and may decay in one of three main possible ways: either it emits positive or negative electrons (β-particles) and so changes the neutron/proton ratio in the direction of stability again, or it achieves the same result by α-emission or by spontaneous fission. *Figure 50* is a more elaborate version of the N–Z chart pictured in *Fig. 6*. It shows the main areas of the Periodic Table in which various types of decay predominate. Note that as we move further and further from stability the added protons or neutrons become less and less strongly bound (*i.e.* the nuclei become

FIG. 50. Chart showing stable and unstable nuclides as a function of N and Z. The black dot marked $^{298}114$ indicates the position of this nuclide which is the possible 'super heavy' nuclide having the two closed shells with 114 protons and 184 neutrons. (Reprinted, by permission, from *Why and how we should investigate nuclides far off the stability line*, G. N. Flerov. Stockholm: Almqvist and Wiksdell, 1967.) (sf = spontaneous fission)

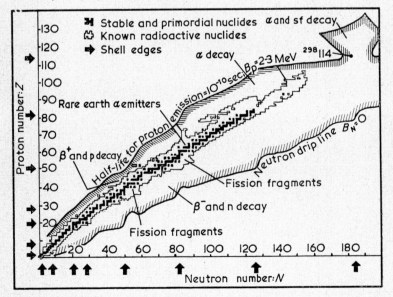

Table 9. Cosmic abundances of some of the elements

Element	Solar	Chondrites type Cc_1	Chondrites type Cc_2	Lunar	Suess & Urey	Goles	Cameron
H	4.8×10^{10}	5.5×10^{6}	3.0×10^{6}	—	4.0×10^{10}	4.8×10^{10}	2.6×10^{10}
Li	1.7	50	16	107	100	16	45
Be	11	—	0.81	—	20	0.81	0.69
C	1.7×10^{7}	8.2×10^{5}	4.5×10^{5}	1000	3.5×10^{6}	1.7×10^{7}	1.35×10^{7}
N	4.6×10^{6}	4.9×10^{4}	2.6×10^{4}	667	6.6×10^{6}	4.6×10^{6}	2.44×10^{6}
O	4.4×10^{7}	7.7×10^{6}	5.5×10^{6}	1.3×10^{6}	2.15×10^{7}	4.4×10^{7}	2.36×10^{7}
Na	9.1×10^{4}	6.0×10^{4}	3.5×10^{4}	2.3×10^{4}	4.38×10^{4}	3.5×10^{4}	6.32×10^{4}
Mg	7.4×10^{5}	1.07×10^{6}	1.04×10^{6}	3.3×10^{5}	9.12×10^{5}	1.04×10^{6}	1.05×10^{6}
Al	6.9×10^{4}	8.5×10^{4}	8.4×10^{4}	4×10^{5}	9.48×10^{4}	8.4×10^{4}	8.51×10^{4}
Si	10^{6}	10^{6}	10^{6}	10^{6}	10^{6}	10^{6}	10^{6}
P	1.9×10^{5}	1.27×10^{4}	8100	4600	1.0×10^{4}	8100	1.27×10^{4}
S	8.0×10^{5}	5.1×10^{5}	2.3×10^{5}	1.3×10^{4}	3.75×10^{5}	8.0×10^{5}	5.05×10^{5}
K	2200	3200	2100	1.0×10^{4}	3160	2100	3240
Ca	6.0×10^{4}	7.2×10^{4}	7.2×10^{4}	5.7×10^{5}	4.90×10^{4}	7.2×10^{4}	7.36×10^{4}
Sc	30	31	35	580	28	35	33
Ti	1800	2300	2400	4.3×10^{5}	2240	2400	2300
V	630	298	590	500	220	590	900
Cr	3800	1.27×10^{4}	1.24×10^{4}	2.0×10^{4}	7800	1.24×10^{4}	1.24×10^{4}
Mn	3000	9300	6200	1.3×10^{4}	6850	6200	8800
Fe	2.5×10^{5}	9.0×10^{5}	8.3×10^{5}	7.3×10^{5}	6.00×10^{5}	2.5×10^{5}	8.90×10^{5}
Co	2400	2200	1900	230	1800	1900	2300
Ni	2.3×10^{4}	4.9×10^{4}	4.5×10^{4}	9.8	2.74×10^{4}	4.5×10^{4}	4.57×10^{4}
Cu	160	590	420	46	212	420	919
Zn	250	1500	630	120	486	630	1500
Ga	20	46	28	47	11.4	28	45.5
Ge	16	130	76	1.7	50.4	76	126

Rb	10	6.0	4.1	30	6.5	4.1	5.95
Sr	25	24	25	1300	18.9	25	58.4
Y	80(?)	4.6	4.7	1000	8.9	4.7	4.6
Zr	20	32	23	3300	54.5	23	30
Nb	10	—	—	230	1.00	0.90	1.15
Mo	10	—	—	5.3	2.42	2.5	2.52
Ru	3	1.85	1.83	—	1.49	1.83	1.6
Rh	1	—	—	—	0.214	0.33	0.33
Pd	—	1.28	1.33	0.057	0.675	1.33	1.5
Ag	0.4	0.95	0.33	0.063	0.26	0.33	0.5
Cd	3	2.1	1.2	0.047	0.89	1.2	2.12
In	1	0.22	0.10	—	0.11	0.10	0.217
Sn	6(?)	4.2	1.7	6.0	1.33	1.7	4.22
Sb	0.1(?)	0.40	0.20	0.060	0.246	0.20	0.381
Ba	16	4.7	5.0	1860	3.66	5.0	4.7
Yb	8(?)	0.21	0.22	—	0.220	0.22	0.21
Pb	4	2.9	1.3	10.7	0.47	1.3	2.90

Notes:

1. All abundance values are normalized to a value of 10^6 atoms for Si (*i.e.* they are cosmic abundance units).
2. The Solar, Chondrite type Cc_1, Chondrite type Cc_3 and Lunar values are averages of measurements on these subjects.
3. The last three columns are evaluated figures by the following authors:

  * H. E. Suess and L. C. Urey, *Rev. Mod. Phys.*, 1956, **28**, 53.

  G. G. Goles, *Handbook of geochemistry*, Vol. 1 (K. H. Wedepohl ed.). Berlin: Springer-Verlag, 1969.

  A. G. W. Cameron in *Origin and distribution of the elements* (L. H. Ahrens ed.). Oxford: Pergamon, 1968.

4. The solar and chondrite values have been taken from the article by G. G. Goles mentioned above: the lunar values have been derived from data given by G. M. Brown, *Endeavour*, 1971, **30**, 147, and refer to measurements made on samples of basalts obtained from the Apollo 11 flight.

  * These values are the ones used in *Fig. 48*.

less stable) until eventually in the case of neutron excessive nuclei, the 'neutron drip line' is reached. This is the point beyond which an extra neutron is not bound at all and hence the nucleus cannot exist. A similar line for protons is not shown in the figure, but the line marked 'half-life for proton emission . . .' indicates that the proton rich nuclei have become very unstable although the binding energy, B_p, of extra protons has not yet reached zero.

Another important feature of the figure is the black arrows marked 'shell edges'. These are at the neutron or proton numbers at which a nucleon shell is filled (magic numbers, p 12) and nuclei having these numbers of nucleons have an enhanced stability, the more so if both proton and neutron shells happen to be closed in the same nucleus. This point will be returned to later when discussing the super-heavy elements (p 109).

ii Why it is possible for nuclei to be stable

In very general terms a nucleus is unstable if there is some other possible arrangement of its components that has a lower total energy (or mass). It is therefore important to be able to calculate the nuclear masses accurately, and when this was originally done by means of the Liquid Drop model (p 11) it was found that they varied smoothly with A and that the calculated values were in reasonable agreement with the measured ones. However, as the measurements became more accurate it became evident that there are deviations from this smooth line which correlate with known nuclear shell structure. Originally the Shell model (p 11) applied only to spherical nuclei, but the later versions take into account the overwhelming evidence that many nuclei are spheroidal. It is now possible to use the Shell model to calculate small corrections to the smooth nuclear masses of the Liquid Drop model.

As far as β-stability is concerned, the picture is quite simple. There is no change in mass number in β-decay and the stable nuclei for any particular value of A are those at the bottom of the mass parabola as described earlier (p 15). Nuclei not at the bottom decay by either β^+ or β^- decay so as to move nearer to it.

Stability against spontaneous fission is a little more difficult to understand. On the simple criterion of lower total mass of the products all nuclei having $A \geq 85$ should decay by spontaneous fission. In fact, the latter, while becoming more prevalent as A increases, is still very improbable even for the lower actinides, although half-lives do start to drop quite rapidly towards the top end of the Periodic Table, until around $Z = 105$ we seem to have reached the limit of nuclear species. The reason for stability existing in spite of the fact that fission would lead to products of lower mass is, as explained earlier (p 36), because as the nucleus stretches out towards the cigar shape it must have before it can break in two, the short range nuclear forces, behaving like a surface tension, oppose the elongation. This results in the energy of the nucleus varying with deformation and giving rise to a potential barrier which inhibits the spontaneous fission. A very similar explanation accounts for α-stability which will not be discussed further.

Calculations based on the simple Liquid Drop model show the barrier to be shaped as in the left-hand diagram of *Fig. 21* and do give a broad outline of the behaviour of spontaneous fission half-lives, but there are so many detailed deviations from measured values that these calculations clearly do not tell the whole story. When spheroidal Shell model effects are taken into account, however, the correlations become very good indeed. Furthermore, the calculations predict that near to the next closed shells ($Z = 114$, $N = 184$) the fission barriers again become higher and so spontaneous fission half-lives increase, giving rise to a new 'island of stability'—the super-heavy nuclei (p 109).

The shape of the fission barrier found from these new calculations is now similar to that shown in the right-hand diagram of *Fig. 21* and this has enabled an explanation to be given for another interesting phenomenon. In 1962 a new and unusual type of spontaneous fission was discovered in which the half-lives were too short (*i.e.* the nuclei were less stable than they should be) by many orders of magnitude if they were calculated by the old methods. These nuclei are called spontaneous fission isomers and are discussed later (p 105).

To summarize, for any nucleus there is a potential barrier against spontaneous fission whose effectiveness depends on its shape, and a similar, though different, barrier against α-decay. These barriers are so high at the lower end of the Periodic Table that the nuclei there can be considered to be stable against these modes of decay. Superimposed on this is the β-stability valley, and so a nucleus may be perfectly stable against β-decay (if it is at the bottom of the valley) but unstable against spontaneous fission, or α-decay, and *vice versa*. The whole pattern of the nuclides found in nature is determined by the past and present interplay of these stability effects.

Actinides, trans-actinides and super-actinides
The actinides are a group of elements whose name arises out of an analogy with the lanthanides. Just as the latter are the group of elements starting at cerium—whose properties bear a marked resemblance to those of lanthanum, so the former start at thorium and resemble actinium. In being the cause of this resemblance the $5f$ electron shell plays the same part with the actinides as the $4f$ shell does with the lanthanides. Carrying the analogy further, the

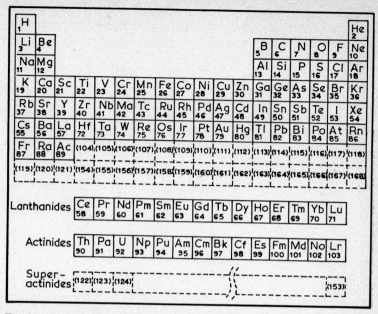

Fig. 51. Expanded Periodic Table showing the 'super-actinides' from element 122 onwards. The 'trans-actinides' are the elements from $Z = 104$ to $Z = 120$ and include the 'super heavy elements' in the region of $Z = 110$–114. (Reprinted, by permission, from 'Elements beyond 100, present status and future prospects', G. T. Seaborg, *A. Rev. nucl. Sci.*, 1968, **18**, 53.)

elements from $Z = 122$ onwards might again form such a group resembling element 121, with the filling of the $6f$ shell, and they have been dubbed the super-actinides. This leaves a gap from $Z = 104$ to 120 in which the elements are expected to occupy their normal places in the Periodic Table. They are usually called the trans-actinides. The expanded Periodic Table is shown in *Fig. 51*.

All the actinides and trans-actinides (and, no doubt, super-actinides) are unstable against some kind of radioactive decay. As shown (p 15), β-stability is a separate issue dependent on the neutron to proton ratio for any particular mass number, but if only stability against α- or spontaneous fission-decay is considered, a general trend can be discerned. From being comparatively stable at thorium, the actinides become more and more short-lived until half-lives in the region above $Z = 102$ are usually only of the order of seconds. However (*see* p 103), an increase of stability is predicted for the transactinide elements near to $Z = 114$, and considerable research effort is now being expended in attempting to prepare them.

The particular property of the actinides which has made them of such paramount importance in the world today, is, of course, the fact that they can easily be induced to undergo nuclear fission (p 35): the world's resources of fossil fuel are running out at an alarming rate and fission is an alternative way of providing sufficient power between the time they do so and the advent of thermonuclear power. This fact, together with their military importance, has meant that very thorough investigations of the properties of the lower actinides (the ones of greatest value in these respects) have been made, and this has in turn set in train a great deal of fundamental research into the preparation and properties of the higher members of the group. Because the information about the actinides is so voluminous only a short summary of some of their more important properties can be given here, in Table 10.

Spontaneous fission isomers

i General description

In 1962, after bombardment of ^{238}U with ^{16}O and ^{22}Ne, a nucleus (later identified as an isomer of ^{242}Am) was discovered (p 103) which decayed by spontaneous fission with a half-life many orders of magnitude too short to fit in with values calculated for the ground states of any of the nuclei which could have been formed. At first it was thought that what was being observed was the spontaneous fission decay of an excited state of one of these nuclei, but it was soon realized that the half-life for γ-decay to the ground state was far too long for this to be the case. It was then suggested that the isomer had a very high nuclear angular momentum, since this would hinder γ-decay (p 28), but measurements soon showed that this was not true either.

As mentioned earlier (p 103) new calculations have led to the realization that the fission barrier is much more complex than it was once thought to be and, for heavy nuclei, has a shape similar to that on the right of *Fig. 21*. It is now believed that a spontaneous fission isomer is a nucleus stretched out into the configuration represented by the secondary minimum in this figure. γ-emission to the normal ground state (*i.e.* the primary minimum in the figure) is hindered because a large change of shape is involved and so the nucleus may decay by spontaneous fission with a half-life controlled by the height and width of the second hump in the fission barrier. The isomer may be formed in the ground state in the secondary minimum, or may be an excited state of that minimum.

ii Experimental methods

The spontaneous fission isomer with the longest known half-life is the one which was discovered first, ^{242}Am with 13 ms, while the shortest half-lives so far identified are in the region of a few nanoseconds. The isomers are readily produced by bombarding the

Table 10. Some properties of the actinides and trans-actinides

Atomic number	Symbol	Name	Gaseous atom electronic configuration (Radon core + . . .)	Density of α-phase of metal in g cm⁻³	Melting point of metal °C	Solution oxidation states	Half-life and main mode of decay of longest lived or most readily available isotope	Typical reaction used in preparation
89†	Ac	Actinium	$6d7s^2$	11.724	1100	+3	^{227}Ac : α : 21.6 y	Natural
90†	Th	Thorium	$6d^27s^2$	15.37	1750	+4	^{232}Th : α : 1.41 × 10¹⁰ y	Natural
91†	Pa	Protactinium	$5f^26d7s^2$	19.07	1565	+4, +5	^{231}Pa : α : 3.23 × 10⁴ y	Natural
92†	U	Uranium	$5f^36d7s^2$	20.45	1132	+3, +4, +5, +6	^{238}U : α : 4.51 × 10⁹ y	Natural
93†	Np	Neptunium	$5f^46d7s^2$	19.82	637	+3, +4, +5, +6, +7	^{237}Np : α : 2.17 × 10⁶ y	β⁻-decay of ^{237}U
94†	Pu	Plutonium	$5f^67s^2$		639.5	+3, +4, +5, +6, +7	^{239}Pu : α : 2.44 × 10⁴ y	^{238}U(n, γ) + 2β⁻-decays
							^{242}Pu : α : 3.79 × 10⁵ y	^{241}Pu(n, γ)
95	Am	Americium	$5f^77s^2$	13.67	995	+3, +4, +5, +6	^{241}Am : α : 458 y	^{241}Pu(n, γ) + β⁻-decay
							^{243}Am : α : 7.97 × 10³ y	^{242}Am(n, γ) + β⁻-decay
96	Cm	Curium	$5f^76d7s^2$	13.51	1340	+3, +4	^{242}Cm : α : 162.5d	^{241}Am(n, γ) + β⁻-decay
							^{244}Cm : α : 17.6 y	^{243}Cm(n, γ)
97	Bk	Berkelium	$5f^86d7s^2$			+3, +4	^{247}Bk : β⁻ : 1.4 × 10³ y	^{244}Cm(α, p)
							^{249}Bk : β⁻ : 314d	^{239}Pu multiple neutron capture + β⁻-decay
98	Cf	Californium	$5f^{10}7s^2$			+3, +5	^{249}Cf : α : 360 y	^{249}Bk β⁻-decay
							^{252}Cf : α : 2.65 y	^{239}Pu multiple neutron capture + β⁻-decay
99	Es	Einsteinium	$5f^{11}7s^2$			+3	^{254}Es : α : ∼300 d	^{239}Pu multiple neutron capture + β⁻-decay
100	Fm	Fermium	$5f^{12}7s^2$			+3	^{257}Fm : α : ∼85 d	^{239}Pu multiple neutron capture + β⁻-decay
101	Md	Mendelevium	$5f^{13}7s^2$			+3	^{256}Md : ec : 90 min	^{253}Es(α, n)
102	No	Nobelium	$5f^{13}7s^2$			+2	^{255}No : α : 180 s	^{246}Cm(^{13}C, 4n)
103	Lr	Lawrencium	$5f^{14}6d7s^2$			? +3	^{256}Lr : α : 35 s	^{243}Am(^{18}O, 5n)

Start of trans-actinides

Atomic number	Symbol	Name	Gaseous atom electronic configuration (Radon core + . . .)	Density	Melting point	Oxidation states	Half-life	Typical reaction
104	Ku* / Rf	Kurchatovium / Rutherfordium	$5f^{14}6d^27s^2$				^{259}Ku : α : 3 s	^{249}Cf(^{13}C, 3n)
105	Ha	Hahnium	$5f^{14}6d^37s^2$				^{261}Ha : α : ≈1 s	^{243}Am(^{22}Ne, 4n)

† = One or more isotopes of each of these elements are found in nature, either as primordal material or as part of a radioactive decay series.

* = Name is disputed and choice depends on whether Russian or American work is finally granted precedence of discovery.

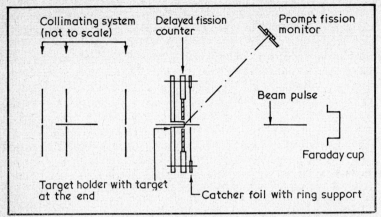

Fig. 52. Apparatus for searching for spontaneous fission isomers. The delayed fission counter cannot 'see' prompt fissions taking place in the target, even if it is switched on during a beam pulse. Spontaneously fissioning species recoil on to the catcher foil and fission there. (Reprinted, by permission, from S. M. Polikanov and G. Sletten, *Nucl. Phys.*, 1970, **A151**, 656.)

appropriate heavy nucleus with either charged particles or neutrons and when one is found the experimenter has three main objectives: (*a*) to prove that a spontaneous fission isomer has been produced, (*b*) to measure its half-life, (*c*) to identify it, *i.e.* to assign its charge and mass numbers. One difficulty which arises is that the reactions which produce the isomers always give rise to a very large amount of

Fig. 53. Apparatus for obtaining the half-life of spontaneous fission emitters. The emitters recoil from the target in the direction of the Faraday cup. If fission occurs the position of the fragments in the track detector material is a measure of the half-life. (Reprinted, by permission, from S. Bjornholm and G. Sletten, *Nucl. Phys.*, 1969, **A139**, 481.)

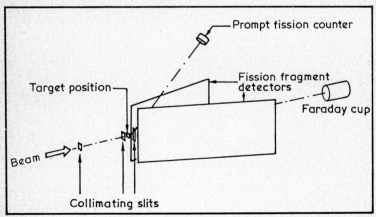

prompt fission, and so the mere observation that fission fragments are being emitted by the target is totally insufficient: it is also necessary to prove that the emission takes place some time after the reaction has occurred. This may be done either by ensuring that the geometry is such that the fission fragment detector cannot respond to prompt fissions, as in *Fig. 52*, or by pulsing the beam of bombarding particles and only switching on the detector in the intervals when the beam is off. If either of these methods is used, the mere observation of delayed fission is enough to achieve objective (*a*).

Objective (*b*) can broadly be reached in three ways. The recoil nucleus itself can start a timing device by means of the emission of an associated particle, *e.g.* in (d, pf) reactions. Timing is stopped when the fission fragment hits the detector. Alternatively, the timing can

Table 11. Spontaneous fission isomers identified up to November 1971

Element	Mass no. (of isomer)	Typical production reaction	Half-life (ns)*
U	236	^{235}U(n, γ)	72; 119
Np	239	^{238}U(α, n)	300
Pu	235	^{233}U(α, 2n)	25
	236	^{237}Np(p, 2n)	<4; 34
	237	^{235}U(α, 2n)	<50; 96; 1010
	238	^{238}U(α, 4n)	<2; 6.5
	239	^{239}Pu(d, pn)	7800
	240	^{239}Pu(d, p)	5.7; 29
	241	^{240}Pu(d, p)	30; 2.7×10^4
	242	^{241}Pu(d, p)	50
	243	^{242}Pu(d, p)	60
Am	237	^{238}Pu(p, 2n)	5
	238	^{239}Pu(p, 2n)	60
	239	^{239}Pu(d, 2n)	155
	240	^{240}Pu(p, n)	9.1×10^6
	241	^{241}Pu(d, 2n)	1500
	242	^{241}Am(n, γ)	14.0×10^6
	243	^{243}Am(d, pn)	6.5×10^3
	244	^{243}Am(n, γ)	8.5×10^5
Cm	241	^{239}Pu(α, 2n)	18
	242	^{240}Pu(α, 2n)	180
	243	^{241}Am(^{3}He, p)	41
	244	^{242}Pu(α, 2n)	>100
	245	^{242}Pu(α, n)	17
Bk	242	^{241}Am(α, 3n)	7.5; >50

* Mean value of all reported data. Note that where two or more values are given this may be caused by one of three factors: (*a*) Half-life measurements are in error. (*b*) Mass or charge assignments are in error. (*c*) Half-lives and assignments are correct and represent several possible states of the second minimum.
(Data supplied by A. T. G. Ferguson, AERE.)

be started by the switching off of the beam, the fragment hitting the detector again stopping it. Finally, the material being bombarded can be mounted on a thin backing so that the compound nucleus recoils out of it, the distance it travels before fissioning then being a measure of the half-life. This is illustrated in *Fig. 53*.

Detection of fission fragments can be by some form of ionization device, such as a gold surface barrier detector (p 76) which is complicated, but which gives the information immediately, or by track detectors (p 76) which are very simple, but involve the experimenter in a large amount of rather tedious scanning after the bombardment is over. An experimental arrangement using track detectors is shown schematically in *Fig. 53*.

It is frequently difficult to achieve objective (c) because, except where the isomers are produced by low energy neutron bombardment, there are usually several possible nuclear reactions which can take place and so give rise to several possible compound nuclei. The assignments usually depend on an application of nuclear systematics and the methods are too diverse to discuss here.

iii *The isomers*

Table 11 is a list of the spontaneous fission isomers which have been identified so far.

Super-heavy elements

i *Probability of their existence*

Super-heavy elements is an indefinite term referring to a group of elements in the trans-actinide region (p 103 and *Fig. 51*) which are predicted to have a much greater stability than the highest actinides. This enhanced stability is due to the presence of the nucleon shells (p 11) which are filled at $Z = 114$ and $N = 184$, giving rise to the 'doubly magic' nucleus $^{298}114$. As far as β-decay is concerned, this nucleus is at the bottom of the β-stability valley (p 15) and the new calculations on fission barriers (p 103) based on the above closed

Table 12. Calculated α- and sf-half-lives for super-heavy nuclides

Z	N	A	$t_{1/2}\alpha$	$t_{1/2}$ sf
112	182	294	10^3 y	10^6 y
	184	296	10^4 y	10^{13} y
	186	298	1 y	10^{13} y
114	182	296	1 y	10^9 y
	184	298	10 y	10^{16} y
	186	300	1 d	10^{15} y
116	182	298	10 s	10^5 y
	184	300	30 s	10^{11} y
	186	302	0.1 s	10^{11} y

shell numbers have shown that a substantial number of nuclides in the region will have quite long spontaneous fission half-lives. Similar calculations on the α-decay barriers show that these are quite high also. Table 12 shows some of the half-life values predicted by S. G. Nilsson.

ii *Possible methods of production*

Since the predictions show the super-heavy nuclides to constitute a sort of island of stability centred around the nuclide $^{298}114$, the problem to be solved is how to get to this island when it is separated from the 'mainland' (ending somewhere in the actinides), by a sea of nuclides having very short half-lives.

There are basically three possible ways in which the island can be reached: (*i*) Neutron bombardment. (*ii*) Bombardment of heavy nuclei by medium mass projectiles. (*iii*) Bombardment of heavy nuclei by heavy mass projectiles.

Neutron bombardment relies on the target nucleus adding on neutrons in a series of (n, γ) reactions, with β-decay competing at every stage along the chain. The general scheme is illustrated in *Fig. 54*. This chain method is the one by which heavy elements are formed in stars (p 93) and it has been successfully used in the artificial production of many of the actinides. However, for it to be successful,

FIG. 54. How neutron bombardment can lead to production of heavy nuclides. There is a competition between (n, γ) reactions and β-decay at each stage, the products depending on the neutron flux, the (n, γ) cross-sections, and the β-decay half-lives.

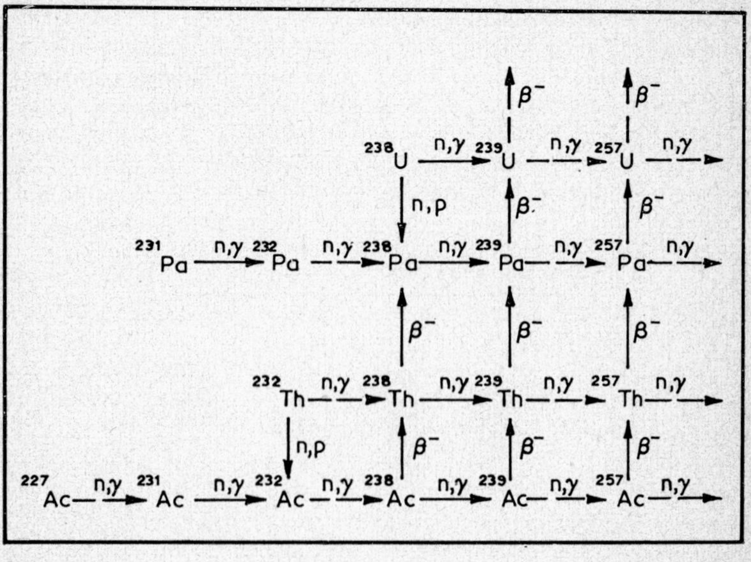

neutron capture, and β-decay must be reasonably evenly balanced so that heavier and heavier nuclei are produced, together with a gradual increase in atomic number. If a region of the N–Z chart is reached where β-half-lives are excessively short, this must be counter-balanced by a proportionately higher neutron flux to increase the probability of the (n, γ) reactions. Typical neutron fluxes are shown in Table 13, and from these values it seems probable that only the stellar r-process is likely to be successful, in view of the extremely short β-half-lives expected in the high actinide and low trans-actinide region.

Table 13. Examples of neutron fluxes available

Neutron producer	flux in $n\ cm^{-2}\ s^{-1}$	Duration of irradiation	Total neutron dose in $n\ cm^{-2}$
High flux reactor	6×10^{15}	A few years	10^{23}
Nuclear explosion	10^{31}	$\sim 1\ \mu s$	10^{25}
Stellar s-process	10^{16}	$\sim 10^3\ y$	10^{26}
Stellar r-process	$> 10^{27}$	1–1000 s	$> 10^{27}$

In the second possible method a target is bombarded by a projectile, the two nuclides being chosen so that the compound nucleus resulting from their fusion is a nucleus on the stable island. The difficulty with this approach is that, because of the way in which the general stability line leans towards more neutron-rich nuclei as Z increases, it is almost impossible to get sufficient neutrons into the product nucleus, and it is therefore on the neutron-poor side of the stable island. The difficulties are further increased by the fact that so much energy must be given to the projectile, for it to be able to overcome the coulomb barrier between it and the target nucleus, (p 32) that the resulting compound nucleus is formed with an excess of energy which it loses by emitting neutrons, thus worsening the situation even more. Typical reactions which might be employed are:

$$^{248}Cm + {}^{40}Ar \rightarrow {}^{284}114 + 4n \qquad \qquad 86$$

$$^{244}Pu + {}^{48}Ca \rightarrow {}^{288}114 + 4n \qquad \qquad 87$$

This method may be more likely to succeed if the target and projectile are chosen so that the compound nucleus formed has a mass of around $A = 320$ and then emits light particles until it reaches the stable island.

The final possibility and one which is at present believed to have a reasonable chance of success, requires that a heavy nucleus is bombarded by a very heavy projectile, e.g.:

$$^{238}U + {}^{238}U \rightarrow {}^{476}184^* \xrightarrow{\text{fission}} {}^{298}114 + {}^{170}Yb + 8n \qquad 88$$

$$^{238}U + {}^{136}Xe \rightarrow {}^{374}146^* \xrightarrow{\text{fission}} {}^{298}114 + {}^{72}Ge + 4n \qquad 89$$

In this case the very heavy compound nucleus first formed fissions and gives, amongst the fission products, various nuclei on the stable island. The problem, of course, is to accelerate to a sufficiently high energy the very heavy projectiles needed, but the new generation of accelerators just (1971–72) coming into use in Russia, the US and France, are able to do this.

A theoretical objection to this approach which has recently been made is that the fission products themselves may fission immediately due to their highly distorted shape when formed.

iii *Identification of the super-heavy elements*
Only a few atoms of the super-heavy nuclides are likely to be made in the first successful experiments and the problem of proving that they are there is a difficult one. It is, in fact, much more difficult than in the case of the higher actinides because these have far more predictable chemical and nuclear properties and their method of production restricts the number of possible products.

Chemistry may be of some help, however, because the elements will fall into their proper places in the Periodic Table (*Fig. 51*, p 104) and so, for example, element 114 should resemble lead in its properties. If, therefore, a substance has been made which has some recognizable nuclear property, such as spontaneous fission or α-decay, this property can be used as a marker to see if it follows the chemistry of lead.

Again, the nuclear properties themselves can be measured, both to eliminate known radioactive substances and to see whether the unknown substance follows the predictions for particular super-heavy nuclides. In this connection fission properties will be especially valuable since these nuclides are expected to have fission fragments with kinetic energies about 20 per cent higher than most of the actinides and are also expected to emit many more prompt neutrons with each fission.

Another possibility is to try to fix the mass number of the unknown nuclide. Provided it emits some radiation by which it can be identified, this could be done by passing it through a mass separator, thus giving the mass directly. Alternatively, if it was an α-emitter and a thin enough source of it could be made, a measurement of the α-decay energy and the energy of the recoiling daughter product nucleus will also give the mass (p 24).

Lastly, it might be possible to fix the atomic number directly by measuring characteristic x-rays (p 53) which might be internally or externally excited.

Probably most of the above methods will have to be applied to the unknown substance, final proof coming by an assembly of all the pieces of the jigsaw into one convincing picture.

iv *Experiments which have been carried out so far*

Experimental work on super-heavy nuclides is really only just beginning. What has been done so far can be roughly divided into examination of possible natural sources of the elements and attempts to make them artificially.

The nuclides might be present in natural sources either because they are primordial material with very long half-lives, or because they are continuously being made in supernova explosions and are arriving here in the cosmic rays. Attempts have been made to find traces of unknown spontaneous fission emitters in old lead and in platinum ore, with results which have been either negative or inconclusive. Cosmic ray plates flown in the stratosphere by balloons have also been examined and there seems to be some evidence of tracks in them which could be due to very heavy nuclei, with atomic numbers close to 110, but again the evidence is conflicting.

As far as artificial attempts are concerned, the neutron route has been tried by exposing targets to atomic bomb explosions, with negative results. The second method, *i.e.* bombardment of heavy nuclei by medium mass projectiles has also failed. However, as shown previously, neither of these approaches ever had much chance of success.

The hopes of the future are pinned on the new accelerators which will be capable of accelerating a large variety of heavy nuclei and so will be able to try out a great many of the variants of the second and third possible production methods. An attempt at this by a back-door route has been made already by a team from the Rutherford High Energy Laboratory in England. Heavy targets were bombarded with 25 GeV protons with the idea that some of the target nuclei would recoil with an energy in excess of 1 GeV and so could become projectiles themselves to bombard other nuclei in the target. The overall cross-section for this double process is very small and results so far are inconclusive. However, there are still hopes of success by this procedure and the work is continuing.

In conclusion, the theoretical evidence for the existence of an island of quite stable nuclei centred around $^{298}114$ is extremely strong, and the chances of reaching it through the medium of the newest heavy ion accelerators seem to be good.

Suggestions for further reading

Chemical effects of nuclear transformations

J. E. Willard, 'Chemical effects of nuclear transformations', *A. Rev. nucl. Sci.*, 1953, **3**, 193.

J. E. Willard, 'Chemical effects of nuclear transformations', *A. Rev. phys. Chem.*, 1955, **6**, 141.

Proc. 1960 and 1964, *Conf. Chem. effects of nucl. transformations.* Vienna: IAEA, 1961 and 1965.

G. Harbottle, 'Chemical effects of nuclear transformations in inorganic solids', *A. Rev. nucl. Sci.*, 1965, **15**, 89.

'Chemical effects of nuclear transformations' by A. G. Maddock and R. Wolfgang in *Nuclear chemistry*, vol. II (L. Yaffe ed.). New York: Academic, 1968.

Isotope effects

M. Wolfgang, 'Isotope effect', *A. Rev. phys. Chem.*, 1969, **20**, 449.

C. J. Collins and N. S. Bowman (eds), *Isotope effects in chemical reactions*, ACS Monograph. New York: Van Nostrand, 1970.

See also the accounts in:

J. F. Duncan and G. B. Cook, *Isotopes in chemistry.* Oxford: Clarendon, 1968.

H. A. C. McKay, *Principles of radiochemistry.* London: Butterworths, 1971.

How the elements are synthesized in stars

W. A. Fowler, 'The origin of the elements', *Proc. natn. Acad. Sci. USA*, 1964, **52**, 524.

E. M. Burbidge, G. R. Burbidge, W. A. Fowler and F. Hoyle, 'Synthesis of the elements in stars', *Rev. mod. Phys.*, 1957, **29**, 547.

B. Pagel, 'Origins of the elements', *New Scient.*, 1965, 8 Apr., 103.

R. J. Taylor, 'The origin of the elements', *Rep. Prog. Phys.*, 1966, **29.2**, 489.

The abundance of the elements

H. E. Suess and H. C. Urey, 'Abundance of the elements', *Rev. mod. Phys.*, 1956, **28**, 53.

L. H. A. Aller, *The abundance of the elements.* New York: Interscience, 1961.

'A new cosmic abundance table' by A. G. W. Camerson in *Origin and distribution of the elements* (L. H. Ahrens, ed.). London: Pergamon, 1968.

'Cosmic abundances' by G. G. Goles in *Handbook of geochemistry*, vol. 1 (K. H. Wedepohl ed.). Berlin: Springer, 1969.

G. M. Brown, 'Geochemistry of the Moon', *Endeavour*, 1971, **30**, 147.

The stability of matter and the double-humped fission barrier

V. M. Strutinsky, 'Shell effects in nuclear masses and deformation energies', *Nucl. Phys.*, 1967, 420.

V. M. Strutinsky, 'Shells in deformed nuclei', *Nucl. Phys.*, 1968, **A122**, 1.

S. G. Nilsson, *Nuclear structure, fission and super heavy elements.* Lawrence Radiation Laboratory Report UCRL 18355, Rev, 1968.

J. R. Nix, *Theories of nuclear fission and super heavy nuclei.* Los Alamos Scientific Laboratory Report LADC 12488, 1971.

Actinides, trans-actinides and super-actinides

G. T. Seaborg, *Man made transuranium elements.* Englewood Cliffs: Prentice-Hall, 1963.

K. W. Bagnall, 'The transuranium elements', *Sci. Prog.*, 1964, **52**, 66.

L. B. Asprey and R. A. Penneman, 'The chemistry of the actinides', *Chem. Engng News*, 1967, Jul 31, 75.

G. T. Seaborg and A. R. Fritch, 'The synthetic elements I, II, III and IV', *Scient. Am.*, 1950, Apr; 1956, Dec; 1963, Apr; 1969, Apr.

Spontaneous fission isomers

A. T. G. Ferguson, 'Recent advances in the study of fissioning isomers and associated phenomena', *Proc. Int. School Nucl. Reacts. 1970* (A. Corciovei, ed.). Bucharest: Inst. of Atomic Phys., 1971.

Super-heavy elements

G. T. Seaborg, 'Elements beyond 100, present status and future prospects', *A. Rev. nucl. Sci.*, 1968, **18**, 53.

W. Greiner, 'Super-heavy elements: extension of the periodic system', *Umschau*, 1969, **69 (12)**, 365.

G. Hermann and K. E. Seyb, 'The heaviest chemical elements', *Naturwissenschaften*, 1969, **56**, 590.

G. T. Seaborg, 'Prospects for further considerable extension of the Periodic Table', *Isotopes Radiat. Technol.*, 1970, **7**, 251.

Glossary

Accelerator: A device for imparting energy to charged particles by subjecting them to electric or magnetic forces.

Alpha (α) particle: A body consisting of two protons and two neutrons, *i.e.* a helium-4 nucleus: emitted in α-decay.

Atomic number (Z): The number of protons in an atomic nucleus. It is a characteristic of a particular chemical element.

Auger electron: An electron emitted as an alternative to an x-ray when a vacancy in an inner electron shell of an atom, caused by emission of a conversion electron, is filled by transfer of an electron from a higher shell.

Beta (β) particle: An electron, emitted in β-decay.

Binding energy: The total binding energy of a nucleus is the amount of energy required to break it down to its constituent parts.

Bosons: Particles obeying Bose–Einstein statistics, *e.g.* photons, α-particles, π-mesons.

Bremsstrahlung: Electromagnetic radiation emitted when an electron is accelerated in the electric field of an atomic nucleus. Used to be called continuous x-rays.

Compton effect: Process in which a γ-ray loses some energy and is deflected by collision with an electron.

Doppler effect: Apparent change in wavelength of a photon caused by the motion of the emitting source.

Electron volt: Energy required to raise an electron through a potential difference of one volt.

Fermi: The nuclear unit of length. One fermi (f) $= 10^{-13}$ cm.

Fermions: Particles obeying Fermi–Dirac statistics *e.g.* electrons, positrons, neutrinos, neutrons, μ-mesons, protons.

Fission: The break-up of a nucleus into two or more large fragments with consequent release of a large amount of energy.

Fusion: The joining together of a number of light particles into one heavier nucleus with consequent release of a large amount of energy.

116

Gamma (γ) ray: Electromagnetic radiation emitted as a result of a reorganization of nucleons in a nucleus.

Half-life: The time taken for the number of radioactive atoms in a sample to be reduced to half its initial value by decay.

Ionization potential: Potential needed to remove an electron from its normal quantum level to infinity.

Ion pair: An electron stripped from an atom by a collision with an energetic particle or ray, together with the positively charged atom also formed in the process.

Isobars: Nuclei having the same mass number (A) but different atomic number (Z).

Isomers: Nuclei having the same mass number (A) and atomic number (Z) but a different level of nuclear excitation.

Isotones: Nuclei having the same neutron number (N) but different atomic number (Z).

Isotopes: Nuclei having the same atomic number (Z) but different mass number (A).

Magic numbers: Neutron or proton numbers in a nucleus which correspond to closed nucleon shells.

Mass number (A): The total number of nucleons in a nucleus.

Mesons: Particles of intermediate mass, *i.e.* between that of the electron and the proton.

Muon: A μ-meson. Its mass is about 207 times that of an electron.

Negatron: A negative electron.

Neutrino: A particle having zero charge and almost zero mass.

Nuclear model: A system, simple enough to submit to mathematical analysis, which is chosen to resemble as closely as possible some aspects of a real system whose properties are incalculable.

Nuclear spin: The angular momentum of a nucleus.

Nucleons: The fundamental constituents of nuclei, *i.e.* the neutrons and protons.

Nuclide: Term referring to an atom having a particular mass number (A) and atomic number (Z).

Pair production: Process in which a γ-ray is completely absorbed in an interaction with the electric field of a nucleus and a positron–negatron pair is created.

Pauli Exclusion Principle: Principle which states that two identical particles cannot occupy a state specified by the same four quantum numbers.

Photoelectric effect: Process in which a γ-ray is completely absorbed and an electron is emitted from an atom.

Photon: Electromagnetic radiation, *e.g.* γ- or x-rays.

Pion: A π-meson.   Its mass is about 270 times that of an electron.

Plasma: A completely ionized gas, consisting of high energy electrons and positive ions.

Positron: A positive electron.

Positronium: A hydrogen-like atom consisting of a positron and a negatron rotating round a common centre of mass.

Thermal neutrons: Neutrons having thermal energies.